"1+X"
大数据应用开发
（Java）
职业技能等级证书系列教材

大数据技术实训教程

预处理、离线分析和实时计算

卢正才 曹小平 郑子伟 ◎ 编著

人民邮电出版社
北京

图书在版编目（CIP）数据

大数据技术实训教程：预处理、离线分析和实时计算 / 卢正才，曹小平，郑子伟编著. -- 北京：人民邮电出版社，2022.10
ISBN 978-7-115-59657-4

Ⅰ. ①大… Ⅱ. ①卢… ②曹… ③郑… Ⅲ. ①数据处理—教材 Ⅳ. ①TP274

中国版本图书馆CIP数据核字(2022)第111452号

内 容 提 要

本书是职业院校大数据相关专业的实训配套教材，也是"1+X"大数据应用开发（Java）职业技能等级证书考试辅助教材。全书共 4 章，包括 Web 服务器日志分析项目、招聘网站数据分析项目、电商网站实时数据分析项目、金融大数据分析项目。本书内容涵盖了大数据技术的完整流程，包括数据采集、数据预处理、数据分析、数据挖掘、数据存储、数据可视化等，既有离线处理，又有实时处理。同时涉及 4 个不同业务背景的 29 个项目任务，包括 17 个职业院校技能大赛项目任务，可帮助读者切实掌握大数据预处理、离线分析和实时计算的实践技能。

本书可作为职业院校、应用型本科院校计算机应用技术、软件技术、软件工程、网络工程和大数据技术等计算机相关专业的教材，还可供从事计算机相关工作的技术人员学习参考。

◆ 编　著　卢正才　曹小平　郑子伟
　　责任编辑　牟桂玲
　　责任印制　王　郁　胡　南

◆ 人民邮电出版社出版发行　北京市丰台区成寿寺路 11 号
　　邮编　100164　电子邮件　315@ptpress.com.cn
　　网址　https://www.ptpress.com.cn
　　北京联兴盛业印刷股份有限公司印刷

◆ 开本：800×1000　1/16
　　印张：15　　　　　　　　　2022 年 10 月第 1 版
　　字数：268 千字　　　　　　2022 年 10 月北京第 1 次印刷

定价：59.90 元

读者服务热线：(010)81055410　印装质量热线：(010)81055316
反盗版热线：(010)81055315
广告经营许可证：京东市监广登字 20170147 号

前言

本书定位

这是一本实训教程,以项目为单元,以需求分析、技术方案设计、过程实现、报告编制为主线,体现了大数据领域的典型任务、典型工作过程和解决方案,旨在提升读者分析解决大数据应用开发常见问题的能力。

本书通过设置不同项目,覆盖了大数据应用开发岗位、职业院校技能大赛大数据技术相关赛项、"1+X"大数据应用开发(Java)职业技能等级证书的要求,是一本"岗课赛证"综合育人的教材。它主要面向职业类院校的广大师生和希望强化大数据应用开发技能的爱好者。

本书主要内容

本书共4章,主要内容是跟随任务步骤完成4个分析项目。

在第1章中,需完成Web服务器日志分析项目,它是"1+X"大数据应用开发(Java)职业技能等级证书(高级)考试的实操真题,是一个典型的大数据离线计算项目。在这个项目中读者将学会如何使用Flume自动采集日志数据,使用MapReduce清洗数据,使用Hive分析数据,使用Sqoop迁移数据,最后利用Java的框架SpringBoot和前端的ECharts库完成数据的可视化。

在第2章中,需完成招聘网站数据分析项目,这也是一个大数据离线计算项目,不同的是它涉及的内容属于职业院校技能大赛大数据技术与应用赛项。在这个项目中,首先需要使用Python合理、合法地"爬取"一些数据,然后完成清洗和分析,最后用Python框架Flask完成数据的可视化并编写数据分析报告。

在第 3 章中，需完成电商网站实时数据分析项目，这是一个代表了工作中常见任务的大数据实时计算项目。在这个项目中，技术栈有较大变化，需要用 Kafka 作为消息中间件实时传递消息，然后使用 Spark 来开展数据分析，利用 Redis 高速缓存结果数据，最后是数据的动态可视化。

在第 4 章中，需完成金融大数据分析项目，其涉及的内容属于职业院校技能大赛大数据技术与应用赛项。相对于第 2 章的项目，本项目的内容更加丰富，技术要求也更加灵活多样，既有离线处理，又有实时处理。本项目中需要使用 Spark 做离线数据分析，使用 Flink 做实时数据计算，使用 Spark ML 做数据挖掘，使用 SpringBoot 和前端技术做可视化，并编写数据分析报告。

本书其他内容

本书是以任务驱动的形式编排内容的，因此主要是项目推进过程的阐述。为便于读者学习，本书还有 3 个重要的配套内容。

第一，知识地图和知识卡片。本书假设读者已经掌握了相关概念和知识，为帮助读者查漏补缺，在每章开头给出了理解该章内容所需的知识地图。受限于篇幅，知识卡片的内容没有出现在教材中，而是整合集结到线上，以数字资源的形式呈现。

第二，环境安装部署。这部分内容很重要，是进行功能开发的基础，但都是按固定顺序操作的命令或动作，且篇幅冗长，如果放在正文中，将打乱读者实现主线功能的思路，因此将其提炼整编到附录中，方便还不熟悉基础环境搭建的读者查阅和练习。

第三，拓展练习。这部分主要是对主线任务的拓展（附加任务）或另外的项目。举一反三有利于巩固技术技能和开阔思路。书中未直接给出拓展练习的答案，而是将其整合到了在线实验环境中，读者可自行挑战这些练习。

使用约定

为了使读者可以更轻松、更有效率地阅读本书，本书在项目任务、程序代码等描述中有一些基本的约定。

• 项目标签

项目标签用于标识本项目是属于"1+X"大数据应用开发（Java）职业技能等级证书（高级）考试中的实操考核项目，还是属于职业院校技能大赛中的大数据技术相关赛项，抑或属于行业应用项目。

• 某网址

程序代码中的"某网址"是代表某网站的地址，在上机操作中，需替换为实际的网址代码，如 https://www.ptpress.com.cn。

配套资源及使用方式

我们为本书配置了丰富的学习和练习资源（网址：https://www.lanqiao.cn/courses/9426），包括：

（1）知识卡片的详细内容；

（2）全书项目的在线实验环境，教师和学生可直接在线上环境里操作；

（3）拓展练习的在线实验环境，教师和学生可直接在线上环境里操作并检验操作的正确性；

（4）微课，即视频讲解。

对于教师，我们还提供额外支持：教师可申请将线上课程引用为私有课，这样可以增补原内容（如上传视频、新增实验、挑战、作业、测试等），也便于以班级为单位把控学生的学习情况。

由于本书编者能力有限，内容难免还有疏漏之处，敬请各位读者批评指正。本书编辑的联系邮箱为 muguiling@ptpress.com.cn。

编者

目 录

第1章 Web 服务器日志分析项目 ···································· 001
- 1.1 任务一：需求分析 ···································· 002
- 1.2 任务二：技术方案设计 ···································· 004
- 1.3 任务三：使用 Flume 采集日志数据 ···································· 006
- 1.4 任务四：使用 MapReduce 清洗数据 ···································· 008
- 1.5 任务五：使用 Hive 分析数据 ···································· 012
- 1.6 任务六：使用 Sqoop 迁移数据 ···································· 016
- 1.7 任务七：Java+ECharts 数据可视化 ···································· 018
- 1.8 答疑解惑 ···································· 039
- 1.9 拓展练习 ···································· 041

第2章 招聘网站数据分析项目 ···································· 043
- 2.1 大赛简介 ···································· 044
- 2.2 任务一：需求分析 ···································· 048
- 2.3 任务二：项目流程 ···································· 050
- 2.4 任务三：使用 Python "爬取" 招聘网站数据 ···································· 051
- 2.5 任务四：使用 MapReduce 预处理数据 ···································· 056
- 2.6 任务五：使用 Hive 分析数据 ···································· 062
- 2.7 任务六：使用 Sqoop 导出数据 ···································· 065
- 2.8 任务七：Flask+ECharts 数据可视化 ···································· 067
- 2.9 任务八：编写分析报告 ···································· 079
- 2.10 答疑解惑 ···································· 080
- 2.11 拓展练习 ···································· 084

第3章 电商网站实时数据分析项目 ···································· 086
- 3.1 任务一：需求分析 ···································· 087

3.2 任务二：项目方案设计 ··· 088
3.3 任务三：使用 Flume+Kafka 实时收集数据 ····························· 089
3.4 任务四：使用 Spark 实时计算数据 ····································· 091
3.5 任务五：Java+ECharts 数据可视化 ···································· 101
3.6 答疑解惑 ·· 113
3.7 拓展练习 ·· 114

第 4 章 金融大数据分析项目 ·· 116

4.1 大赛简介 ·· 117
4.2 任务一：需求分析 ··· 119
4.3 任务二：项目流程 ··· 121
4.4 任务三：使用 Spark 抽取离线数据 ···································· 123
4.5 任务四：使用 Spark 统计离线数据 ···································· 125
4.6 任务五：使用 Flume+Kafka 实时采集数据 ··························· 128
4.7 任务六：使用 Flink 实时计算数据 ···································· 130
4.8 任务七：Vue.js+Java+ECharts 数据可视化 ·························· 136
4.9 任务八：使用 Spark ML 数据挖掘 ···································· 156
4.10 任务九：编写分析报告 ··· 160
4.11 答疑解惑 ··· 162
4.12 拓展练习 ··· 163

附录 ·· 165

附录 1　Hadoop 安装部署和配置 ··· 165
附录 2　掌握 HDFS Shell 操作 ·· 175
附录 3　通过 WordCount 熟悉 MapReduce ······························ 182
附录 4　深入理解 MapReduce ·· 186
附录 5　Flume 安装部署和配置 ··· 199
附录 6　Hive 安装部署和配置 ··· 200
附录 7　Sqoop 安装部署和配置 ··· 201
附录 8　Hadoop 高可用集群环境安装部署和配置 ························ 203
附录 9　Hadoop 集群节点动态管理 ······································· 212
附录 10　Kafka 安装部署和配置 ·· 214
附录 11　Spark 安装部署和配置 ·· 217
附录 12　Spark RDD 算子 ·· 220
附录 13　通过 WordCount 熟悉 Spark RDD ····························· 230
附录 14　Flink 安装部署和配置 ··· 231

第1章 Web 服务器日志分析项目

项目标签:"1+X"证书

【本章简介】

本章以 Web 服务器日志分析项目的推进为主线来介绍大数据离线计算项目的详细操作流程,力求帮助读者掌握大数据离线计算的流程及所需工具、技术。该项目是"1+X"大数据应用开发(Java)职业技能等级证书(高级)考试中的实操考核项目,熟悉本项目有助于证书的考取。

本章首先分析项目的需求和技术架构,帮助读者了解项目目标,厘清大体思路。其次按照大数据离线计算的流程分步骤详细介绍数据采集、数据清洗、数据分析、数据迁移、数据可视化等操作。最后通过答疑解惑和拓展练习帮助读者巩固知识,强化技能。

【知识地图】

本章项目涉及的知识地图如图 1.1 所示,读者在学习过程中可以查询相关知识卡片,补充理论知识。

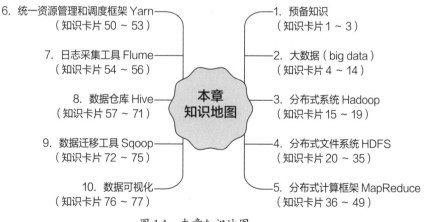

图 1.1 本章知识地图

1.1 任务一：需求分析

1．业务背景

某大型电商公司，其产生的原始日志某小时（任选 1 小时）高达 45303452 条，某天（任选 1 天）日志量为 422110779 条。

用于大数据处理的设备规划如表 1.1 所示。

表 1.1　大数据处理的设备规划

功能	主机型号	数量	CPU	MEM	HDD
采集与解析服务器	浪潮 NF5240M3	3	2×12Core	64GB 内存	2×600GB 15000r/min SAS 硬盘
预处理服务器	浪潮 NF5240M3	8	2×12Core	64GB 内存	6×600GB 10000r/min SAS 硬盘
ETL 服务器 Hadoop 平台	浪潮 NF5240M3	21	2×12Core	64GB 内存	16×1TB 7200r/min SAS 硬盘

已知该网站服务器的日志格式如下所示。

```
2017-06-19 00:26:36 101.200.190.54 GET /sys/ashx/ConfigHandler.ashx action=js
8008 - 60.23.128.118 Mozilla/5.0+(Windows+NT+6.1;+WOW64;+rv:53.0)+Gecko/20100101+
Firefox/53.0 http:// 某网址 /Welcome.aspx 200 0 0 297
```

每一行数据共有 15 个属性值，分别用空格分隔，其含义如下。

第 1 列：日期，如 2017-06-19。

第 2 列：时间，如 00:26:36。

第 3 列：服务器端 IP 地址，如 101.200.190.54。

第 4 列：请求方式，如 GET。

第 5 列：访问路径，不带参数，如 /sys/ashx/ConfigHandler.ashx。

第 6 列：访问参数，如 action=js。

第 7 列：访问端口，如 8008。

第 8 列：访问用户，如 -，表示没有值。

第 9 列：客户端 IP 地址，如 60.23.128.118。

第 10 列：客户端浏览器，如 Mozilla/5.0+(Windows+NT+6.1;+WOW64;+rv:53.0)+

Gecko/2010 0101+Firefox/53.0。

第 11 列：跳转之前的访问路径，如 http:// 某网址 /Welcome.aspx。

第 12 列：页面返回状态码，如 200。

第 13 列：页面返回子状态码，如 0。

第 14 列：操作相应状态码，如 0。

第 15 列：返回页面所需时间，如 297 毫秒。

2．业务需求

（1）企业高层希望了解网站的流量情况，比如流量来源、网站访客特性等，从而帮助提高网站流量，提升网站访客体验，让访客更多地沉淀下来变成会员或客户，通过更少的投入获取最大化的收入。

（2）企业高层期望通过掌握的大数据来了解公司的运行情况，如最受欢迎或最不受欢迎的功能模块。对于所有企业来说，除了要保证网站稳定、正常地运行以外，还要关注细分功能访问量的统计和分析报表，这对于提高服务能力和服务水平是必不可少的。

（3）企业高层期望掌握网站的缺陷情况，以便进行优化。

3．用户需求

现在运营、运维等部门根据业务需求，提出了以下具体需求。

（1）运营部门想知道最活跃的用户有哪些。

（2）运维部门想知道经常被访问但无法响应的页面有哪些。

（3）产品部门想知道最常被访问的页面有哪些，以便进一步加强。

4．功能需求

（1）统计分析客户端访问次数最多的前 10 个 IP 地址，按照访问次数降序排列，并采用 ECharts 柱状图可视化。

（2）分析客户端访问页面中不正常的页面（页面返回状态码不为 200），统计访问次数最多的前 10 个不正常页面的访问路径，按照访问次数降序排列，并采用 ECharts 折线图可视化。

（3）统计分析客户端访问次数最多的前 10 个访问路径，按照访问次数降序排列，并采用 ECharts 折线图可视化。

1.2 任务二：技术方案设计

1．技术领域判定

通过需求分析，我们知道该项目有几个特征。

（1）数据量大。

（2）不需要实时分析。

（3）需要有简单的报表。

这是一个对海量日志进行统计分析的场景，解决方案是典型的大数据离线计算技术。

2．技术选型

大数据离线分析的行业最佳实践将架构由下往上分为 5 层：数据采集层、数据存储层、数据分析层、机器学习层、数据展示层。由于本项目需求不涉及机器学习，因此，我们只需要对其他 4 层进行选型。

（1）数据采集层。我们需要的原始数据分布在不同服务器上，服务器通常会提供数据访问、下载的接口，通过远程的方式来采集。可以使用的技术方案有 Flume、Scribe 等。其中 Flume 是一个纯粹为数据采集而生的分布式服务，我们这里选用 Flume。

（2）数据存储层。使用目前应用最广泛的大数据分析平台 Hadoop 的文件系统 HDFS 来存储数据。

（3）数据分析层。该层技术较多，可以使用的技术有 MapReduce、Hive、Spark、Impala 等。考虑到本项目的数据还需要进行数据清洗（预处理），而 MapReduce 可以操作数据细节，因此这里选择 MapReduce。对实时性要求不高的离线数据分析，既可以用 MapReduce 也可以用 Hive，我们选用支持 SQL 语法的 Hive 以减少代码量。而 Spark 等技术，主要用于实时分析。

（4）数据展示层。展示层技术较多，可根据技术团队的技术基础自行选择，这里我们选择 Java 技术栈里较为主流的 SpringBoot+SSM（由 Spring+SpringMVC+MyBatis 框架整合而成，常作为 Java Web 项目的框架）作为后端技术，结合 ECharts 等前端技术，以图形化的方式来展示数据结果，如饼图、柱状图、折线图等。

3．实现流程

本项目的流程，如图 1.2 所示，先将一部分 Web 服务器日志数据作为原始数据，使用

Flume 采集到 HDFS 上，使用 MapReduce 进行数据清洗，清洗后的数据放到 HDFS 上，然后使用 Hive 进行数据分析，分析结果存放到 HDFS 上，之后使用 Sqoop 将数据分析结果导出到 MySQL 中，再创建 SpringBoot 应用程序来访问数据库，最后使用 ECharts 呈现在网页上。

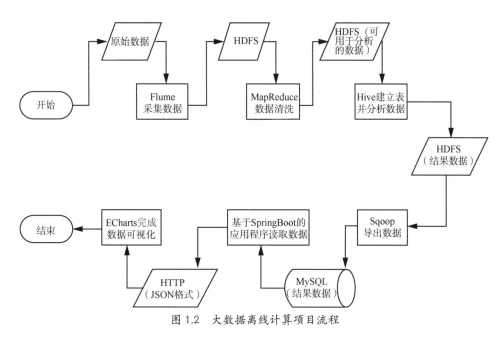

图 1.2　大数据离线计算项目流程

4．开发环境准备

在完成后续任务之前，我们需要安装部署并配置一系列的软件，准备好开发环境，如表 1.2 所示。这些基础操作，我们放到附录中以供不熟悉的同学参考。

表 1.2　开发环境准备

操作	附录
Hadoop 安装部署和配置	附录 1
Flume 安装部署和配置	附录 5
Hive 安装部署和配置	附录 6
Sqoop 安装部署和配置	附录 7

1.3 任务三：使用 Flume 采集日志数据

使用 Flume 采集数据的流程如图 1.3 所示。

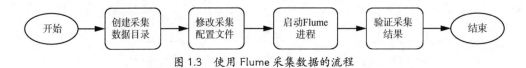

图 1.3 使用 Flume 采集数据的流程

按照操作步骤，可以将 Flume 采集数据的过程划分为 4 个阶段，分别是创建采集数据目录、修改采集配置文件、启动 Flume 进程和验证采集结果。

这里我们将数据直接放到本地一个目录下，作为采集数据目录。Flume 使用 Spooling Directory Source 方式监听一个指定的目录，只要向这个指定的目录中添加新的文件，Source 组件就可以获取该信息，并解析该文件的内容，然后写入 Channel。等 Sink 处理完之后，标记该文件已完成处理，文件名会被添加 ".completed" 后缀。虽然 Source 组件自动监控整个目录，但是只能监控文件，如果向被标记为已完成处理的文件中添加内容，Source 不能识别；如果在被监控目录下出现了一个新文件，则会立即被处理。值得注意的是，无法监控子目录的变动。

下面我们开始准备采集目录 ~/data 下的文件数据。

1. 下载原始数据

获取数据文件：

```
shiyanlou:~/ $ wget https://labfile.oss.aliyuncs.com/courses/2956/a.txt
```

2. 创建采集数据目录

在 ~ 目录下新建目录 data。

```
shiyanlou:~/ $ mkdir data
```

3. 修改采集配置文件

创建 Flume 配置文件，监听 ~/data（/home/shiyanlou/data）目录，并把数据文件 a.txt 采集到 HDFS 上。

在 ~ 目录下新建文件 spool.conf，完整代码如程序清单 1.1 所示。

程序清单 1.1

```
shiyanlou:~/ $ vi spool.conf
a1.sources=r1
a1.sinks=k1
a1.channels=c1

a1.sources.r1.type=spooldir
a1.sources.r1.spoolDir=/home/shiyanlou/data
a1.sources.r1.fileHeader=true

a1.sinks.k1.type=hdfs
a1.sinks.k1.channel=c1
a1.sinks.k1.hdfs.path=/flume/
a1.sinks.k1.hdfs.filePrefix=log-
a1.sinks.k1.hdfs.round=true
a1.sinks.k1.hdfs.roundValue=10
a1.sinks.k1.hdfs.roundUnit=minute
a1.sinks.k1.hdfs.rollInterval=3
#设置每多少个字节上传一次
a1.sinks.k1.hdfs.rollSize=0
#设置每多少条数据上传一次
a1.sinks.k1.hdfs.rollCount=3000
a1.sinks.k1.hdfs.batchSize=1
a1.sinks.k1.hdfs.useLocalTimeStamp=true
#生成的文件类型，默认是 Sequencefile，可用 DataStream，则为普通文本
a1.sinks.k1.hdfs.fileType=DataStream

a1.channels.c1.type=memory
a1.channels.c1.capacity=1000
a1.channels.c1.transactionCapacity=100

a1.sources.r1.channels=c1
a1.sinks.k1.channel=c1
```

参数 a1.sources.r1.type=spooldir 指明了 Flume 采集的方式为 spooldir。

参数 a1.sources.r1.spoolDir=/home/shiyanlou/data 指明了从目录 /home/shiyanlou/data 采集数据。

参数 a1.sinks.k1.type=hdfs 指明了数据将被采集到 HDFS 上。

参数 a1.sinks.k1.hdfs.path=/flume/ 指明了数据将被采集到 HDFS 上的目录 /flume/ 中。

4．启动 Flume 进程

由于这里使用了 Hadoop，所以先启动 Hadoop。

```
shiyanlou:~/ $ start-all.sh
......
```

再启动 Flume。

```
shiyanlou:~/ $ flume-ng agent -n a1 -f /home/shiyanlou/spool.conf -Dflume.root.logger=INFO,console
......
```

启动后会发现这是一个前台进程。

5．验证采集结果

另开一个 Shell 终端，复制数据 a.data 文件到 ~/data 目录中。

```
shiyanlou:~/ $ cp ~/a.data ~/data/
```

然后查看 HDFS 上是否已经有数据文件，并且内容和原始数据一样。

```
shiyanlou:~/ $ hadoop fs -ls /flume
......
```

可以看到文件已经上传，Flume 进程可以终止（使用"Ctrl+C"组合键）。

1.4 任务四：使用 MapReduce 清洗数据

本任务涉及 HDFS Shell 操作和 MapReduce 编程，对此不熟悉的同学可以参考表 1.3 中提到的相关附录进行学习。

表 1.3 HDFS Shell 操作和 MapReduce 编程

操作	附录
掌握 HDFS Shell 操作	附录 2
通过 WordCount 熟悉 MapReduce	附录 3
深入理解 MapReduce	附录 4

使用 MapReduce 清洗数据的流程如图 1.4 所示。

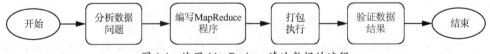

图 1.4 使用 MapReduce 清洗数据的流程

按照操作步骤，可以将 MapReduce 清洗数据的过程划分为 4 个阶段，分别是分析数据问题、编写 MapReduce 程序、打包执行和验证数据结果。

1．分析数据问题

项目中采集的数据可能存在数据缺失、数据格式不良好、数据不正确、数据前后矛盾或不一致等数据质量问题，会影响我们后续的数据分析操作，甚至导致错误。因此我们需要先对原始数据进行一次有效清洗。

通过观察采集到的原始数据，发现主要有如下两个问题。

（1）有些行是以 # 开头的，表示该行为注释内容，需要删除。

（2）有些访问路径是以 xxx.js、xxx.css、xxx.jpg、xxx.png、xxx.gif 等结尾的，它们是伴随着访问页面的同时访问的，会极大地干扰真正访问路径的统计，所以需要把这些行（包含以上面 5 种格式结尾的行）删除。

2．编写 MapReduce 程序

（1）创建 maven 项目。

打开 Eclipse，新建一个 maven project，操作过程如下。

右击项目浏览器 → 新建项目 → 选择 "maven project"。设置 catalog 为 "internal"，项目类型为 "maven-archetype-quickstart"，Group Id 为 "org.lanqiao"，Artifact Id 为 "bigdata"，单击 "finish"。

为 pom.xml 配置本项目对 jar 包的依赖，代码如下。

```xml
<dependencies>
    <dependency>
        <groupId>junit</groupId>
        <artifactId>junit</artifactId>
        <version>4.12</version>
    </dependency>
    <dependency>
        <groupId>log4j</groupId>
        <artifactId>log4j</artifactId>
        <version>1.2.17</version>
    </dependency>
    <dependency>
        <groupId>org.apache.hadoop</groupId>
        <artifactId>hadoop-client</artifactId>
        <version>2.9.2</version>
    </dependency>
    <dependency>
```

```xml
        <groupId>org.apache.hadoop</groupId>
        <artifactId>hadoop-common</artifactId>
        <version>2.9.2</version>
    </dependency>
</dependencies>
```

（2）编写 MapReduce 程序。

新建类 PreProcess，完整的代码如程序清单 1.2 所示。

程序清单 1.2

```java
package org.lanqiao.bigdata;

import java.net.URI;
import java.util.regex.Pattern;

import org.apache.hadoop.conf.Configuration;
import org.apache.hadoop.fs.FileSystem;
import org.apache.hadoop.fs.Path;
import org.apache.hadoop.io.LongWritable;
import org.apache.hadoop.io.NullWritable;
import org.apache.hadoop.io.Text;
import org.apache.hadoop.mapreduce.Job;
import org.apache.hadoop.mapreduce.Mapper;
import org.apache.hadoop.mapreduce.lib.input.FileInputFormat;
import org.apache.hadoop.mapreduce.lib.output.FileOutputFormat;

public class PreProcess {
    public static void main(String[] args) throws Exception {
        Configuration conf=new Configuration();
        // 获取集群访问路径
        FileSystem fileSystem=FileSystem.get(new URI("hdfs://localhost:9000/"),conf);
        // 判断输出路径是否已经存在，如果存在，则删除
        Path outPath=new Path(args[1]);
        if(fileSystem.exists(outPath)){
            fileSystem.delete(outPath,true);
        }
        // 生成job，并指定job 的名称
        Job job=Job.getInstance(conf,"PreProcess");
        // 指定打成jar 后的运行类
        job.setJarByClass(PreProcess.class);
        // 指定mapper 类
```

```
job.setMapperClass(MyMapper.class);
        //指定 mapper 的输出类型
job.setMapOutputKeyClass(Text.class);
job.setMapOutputValueClass(NullWritable.class);
        //无须 Reducer

        //数据的输入目录
FileInputFormat.addInputPath(job,new Path(args[0]));
        //数据的输出目录
FileOutputFormat.setOutputPath(job,new Path(args[1]));
System.exit(job.waitForCompletion(true)?0:1);
    }

    static class MyMapper extends Mapper<LongWritable,Text,Text,NullWritable>{
        protected void map(LongWritable k1,Text v1,Context context) throws java.
io.IOException ,InterruptedException {
            final String s=v1.toString();
            //遇到#开头的行直接舍弃
            if(s.startsWith("#")) {
                return;
            }
            //遇到访问路径是以 xxx.js、xxx.css、xxx.jpg、xxx.png、xxx.gif 等结尾的行
就直接舍弃
            String[] ss=s.split("");
              if(ss[4].endsWith(".js")||ss[4].endsWith(".css")||ss[4].endsWith
(".jpg")
                    ||ss[4].endsWith(".png")||ss[4].endsWith(".gif")) {
                return;
            }
            //返回新的清洗后的一行数据
  context.write(v1,NullWritable.get());
        };
    }
}
```

3．打包执行

程序编写好之后，使用 Eclipse 的 export 来导出 jar 包，可以只选择要导出的类，操作过程如下。

右击项目 → 选择"export" → 选择"java" → 选择"jar file" → 单击"next" → 选中 PreProcess 类 → 设置保存位置为"/home/shiyanlou" → 设置 jar 的名称为"PreProcess.jar"。

然后执行该 jar。

```
shiyanlou:~/ $ hadoop jar PreProcess.jar org.lanqiao.bigdata.PreProcess /flume /out
```

执行完毕后，查看输出结果。

```
shiyanlou:~/ $ hadoop fs -cat /out/part*
  2017-06-19 00:14:44 101.200.190.54 GET / - 8008 - 42.156.251.185 spiderman - 302 0 0 2370
  2017-06-19 00:26:33 101.200.190.54 GET / - 8008 - 60.23.128.118 Mozilla/5.0+(Windows+NT+6.1;+WOW64;+rv:53.0)+Gecko/20100101+Firefox/53.0 - 200 0 0 1372
  2017-06-19 00:26:33 101.200.190.54 GET /sys/ashx/ConfigHandler.ashx action=js 8008 - 60.23.128.118 Mozilla/5.0+(Windows+NT+6.1;+WOW64;+rv:53.0)+Gecko/20100101+Firefox/53.0 http://某网址/ 200 0 0 338
  2017-06-19 00:26:35 101.200.190.54 GET /ashx/MenuData.ashx - 8008 - 60.23.128.118 Mozilla/5.0+(Windows+NT+6.1;+WOW64;+rv:53.0)+Gecko/20100101+Firefox/53.0 http://某网址/ 200 0 0 1063
  ……
```

4．验证数据结果

查看数据结果没有发现以 # 开头的行和访问路径以 xxx.js、xxx.css、xxx.jpg、xxx.png、xxx.gif 等结尾的行，内容符合要求。

1.5 任务五：使用 Hive 分析数据

使用 Hive 分析数据的流程如图 1.5 所示。

图 1.5 使用 Hive 分析数据的流程

按照操作步骤，可以将 Hive 分析数据的过程分为 4 个阶段，分别是导入分析数据、分析需求、编写 HQL 语句和执行并验证结果。

数据清洗之后的数据存储在 HDFS 上的 /out/part* 文件中，下面使用 Hive 来完成数据分析。

针对需求一：统计分析客户端访问次数最多的前 10 个 IP 地址，按照访问次数降序排列，并采用 ECharts 柱状图可视化。

(1)导入分析数据。

开启 Hive 客户端,然后在 Hive 中创建数据库 t1,在 t1 数据库中创建内部表 a1,该表有 15 个字段,分别和数据中的 15 列保持对应,列之间用空格分割,将数据文件 /out/part* 导入该表,代码如下。

```
shiyanlou:~/ hive
hive> create database t1;
hive> use t1;
hive> create table a1(n1 string,n2 string,n3 string,n4 string,n5 string,n6 string,n7 string,n8 string,n9 string,n10 string,n11 string,n12 string,n13 string,n14 string,n15 string) row format delimited fields terminated by " ";
hive> load data inpath 'hdfs://localhost:9000/out/part*' into table a1;
hive> select * from a1 limit 10;
……
```

(2)分析需求。

针对需求一,客户端访问 IP 地址是第 9 列,如 60.23.128.118。如果要统计 IP 地址的访问次数,可以使用 Hive 中的聚合函数 count( ) 来计数,这里分析时应该按 IP 地址分组统计访问次数(一行数据代表访问一次)。在可视化时,再在数据库中降序排列,取前 10 条数据即可。

(3)编写 HQL 语句。

思路:在 HDFS 上创建一个目录,把分析数据以覆盖的方式插入该目录。

在 HDFS 上创建目录。

```
shiyanlou:~/ $ hadoop fs -mkdir /user/hive/warehouse/a1
```

然后把分析数据以覆盖的方式插入该目录。

```
hive>insert overwrite directory 'hdfs://localhost:9000/user/hive/warehouse/a1' select n9,count(*) as num from a1 group by n9;
```

(4)执行并验证结果。

执行完毕之后,可以查看该目录下的文件内容是否正确。

```
shiyanlou:~/ $ hadoop fs -cat /user/hive/warehouse/a1/*
101.226.33.201      1
101.226.61.207      1
101.226.65.102      2
101.226.68.213      3
101.226.89.115      1
101.226.89.121      1
101.226.89.122      1
```

101.226.99.197	1
104.152.52.55	1
104.152.52.56	1
104.152.52.58	1
104.152.52.59	1
114.248.38.224	283
114.248.50.124	26
180.153.163.190	2
180.163.1.46	1
180.163.2.116	1
204.93.180.6	1
42.156.251.185	1
60.23.128.118	74
61.151.226.189	1
61.151.228.22	1

针对需求二：分析客户端访问页面中不正常的页面（页面返回状态码不为200），统计访问次数最多的前10个不正常页面的访问路径，按照访问次数降序排列，并采用ECharts折线图可视化。

（1）导入分析数据。

数据库和表之前已经建好，数据也已经导入完毕。

（2）分析需求。

对于需求二，页面返回状态码是第12列，如200。不正常的页面，其返回状态码不为200，这是过滤条件。访问路径是第5列，这里需要按访问路径分组统计访问次数。在可视化时，再在数据库中降序排列，取前10条数据即可。

（3）编写HQL语句。

思路：在HDFS上创建一个目录，把分析数据以覆盖的方式插入该目录。

在HDFS上创建目录。

```
shiyanlou:~/ $ hadoop fs -mkdir /user/hive/warehouse/a2
```

然后把分析数据以覆盖的方式插入该目录。

```
hive>insert overwrite directory 'hdfs://localhost:9000/user/hive/warehouse/a2'
select n5 as name,count(*) as num from a1 where n12!=200 group by n5;
```

（4）执行并验证结果。

执行完毕之后，可以查看该目录下的文件内容是否正确。

```
shiyanlou:~/ $ hadoop fs -cat /user/hive/warehouse/a2/*
/    12
/BlueVacation/Controller/SecondRecommendHandler.ashx         2
/BlueVacation/Controller/SubShopHandler.ashx           9
/BlueVacation/View/FINANCE/FinancialAccounts/Sheet.aspx      5
/BlueVacation/View/RETAILSALESMGR/Payment/Default.aspx       1
/uploadfile/RetailSales/          2
```

针对需求三：统计分析客户端访问次数最多的前 10 个访问路径，按照访问次数降序排列，并采用 ECharts 折线图可视化。

（1）导入分析数据。

数据库和表之前已经建好，数据也已经导入完毕。

（2）分析需求。

对于需求三，访问路径是第 5 列，这里需要使用 Hive 中的聚合函数 count() 来统计访问路径的次数。在可视化时，再在数据库中降序排列，取前 10 条数据即可。

（3）编写 HQL 语句。

思路：在 HDFS 上创建一个目录，把分析数据以覆盖的方式插入该目录。

在 HDFS 上创建目录。

```
shiyanlou:~/ $ hadoop fs -mkdir /user/hive/warehouse/a3
```

然后把分析数据以覆盖的方式插入该目录。

```
hive>insert overwrite directory 'hdfs://localhost:9000/user/hive/warehouse/a3'
select n5,count(*) as num from a1 group by n5;
```

（4）执行并验证结果。

执行完毕之后，可以查看该目录下的文件内容是否正确。

```
shiyanlou:~/ $ hadoop fs -cat /user/hive/warehouse/a3/*
/    24
/BlueVacation/Controller/ContractApplicationHandler.ashx    1
/BlueVacation/Controller/FinancialAccountsHandler.ashx      5
/BlueVacation/Controller/FirstRecommendHandler.ashx         5
/BlueVacation/Controller/GroupStageHandler.ashx       9
/BlueVacation/Controller/InvoiceHandler.ashx          2
/BlueVacation/Controller/LanEmailHandler.ashx         24
/BlueVacation/Controller/LvyouAreaHandler.ashx        22
/BlueVacation/Controller/NewsHandler.ashx    24
/BlueVacation/Controller/PaymentHandler.ashx          5
/BlueVacation/Controller/ReceivablesHandler.ashx      2
/BlueVacation/Controller/RetailSalesHandler.ashx      25
```

```
/BlueVacation/Controller/SecondRecommendHandler.ashx     7
/BlueVacation/Controller/SubContractHandler.ashx        16
/BlueVacation/Controller/SubInvoiceHandler.ashx          3
/BlueVacation/Controller/SubOrdersHandler.ashx           9
……
```

1.6 任务六：使用 Sqoop 迁移数据

使用 Sqoop 迁移数据的流程如图 1.6 所示。

图 1.6 使用 Sqoop 迁移数据的流程

按照操作步骤，可以将 Sqoop 迁移数据的过程划分为 3 个阶段，分别是建库和建表、编写 Sqoop 命令导出数据及验证数据结果。

针对需求一：统计分析客户端访问次数最多的前 10 个 IP 地址，按照访问次数降序排列，并采用 ECharts 柱状图可视化。

（1）建库和建表。

首先登录 MySQL，笔者的 MySQL 数据库账号为 hive，密码为 hive。然后在 MySQL 中创建数据库 sqoop，再在 sqoop 下创建数据表 a1。a1 含有 2 个字段：一个是 name，字符串类型；另一个是 num，整数类型。

```
shiyanlou:~/ $ mysql -u hive -p
Enter password:
```

然后输入密码"hive"。

```
mysql> create database sqoop;
mysql> use sqoop;
mysql> create table a1(name varchar(100),num int);
```

（2）编写 Sqoop 命令导出数据。

使用 Sqoop 把 HDFS 上的数据导出到 MySQL 中。

```
shiyanlou:~/ $ sqoop export --connect jdbc:mysql://localhost:3306/sqoop
--username hive --password hive --table a1 --export-dir /user/hive/warehouse/a1
--input-fields-terminated-by '\001' --num-mappers 1
```

（3）验证数据结果。

执行完毕后，查看 MySQL 中的数据是否正确。

针对需求二：分析客户端访问页面中不正常的页面（页面返回状态码不为 200），统计访问次数最多的前 10 个不正常页面的访问路径，按照访问次数降序排列，并采用 ECharts 折线图可视化。

（1）建库和建表。

首先登录 MySQL，然后在数据库 sqoop 下创建数据表 a2。a2 含有 2 个字段：一个是 name，字符串类型；另一个是 num，整数类型。

```
shiyanlou:~/ $ mysql -u hive -p
Enter password:
```

输入密码"hive"。

```
mysql> use sqoop;
mysql> create table a2(name varchar(100),num int);
```

（2）编写 Sqoop 命令导出数据。

使用 Sqoop 把 HDFS 上的数据导出到 MySQL 中。

```
shiyanlou:~/ $ sqoop export --connect jdbc:mysql://localhost:3306/sqoop --username hive --password hive --table a2 --export-dir /user/hive/warehouse/a2 --input-fields-terminated-by '\001' --num-mappers 1
```

（3）验证数据结果。

执行完毕后，查看 MySQL 中的数据是否正确。

针对需求三：统计分析客户端访问次数最多的前 10 个访问路径，按照访问次数降序排列，并采用 ECharts 折线图可视化。

（1）建库和建表。

首先登录 MySQL，然后在数据库 sqoop 下创建数据表 a3。a3 含有 2 个字段：一个是 name，字符串类型；另一个是 num，整数类型。

```
shiyanlou:~/ $ mysql -u hive -p
Enter password:
```

输入密码"hive"。

```
mysql> use sqoop;
mysql> create table a3(name varchar(100),num int);
```

（2）编写 Sqoop 命令导出数据。

使用 Sqoop 把 HDFS 上的数据导出到 MySQL 中。

```
shiyanlou:~/ $ sqoop export --connect jdbc:mysql://localhost:3306/sqoop
--username hive --password hive --table a3 --export-dir /user/hive/warehouse/a3
--input-fields-terminated-by '\001' --num-mappers 1
```

（3）验证数据结果。

执行完毕后，查看 MySQL 中的数据是否正确。

1.7 任务七：Java+ECharts 数据可视化

Java+ECharts 数据可视化的流程如图 1.7 所示。

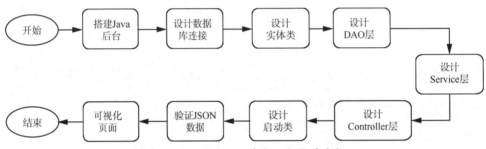

图 1.7　Java+ECharts 数据可视化的流程

按照操作步骤，可以将 Java+ECharts 数据可视化的过程划分为 9 个阶段，分别是搭建 Java 后台、设计数据库连接、设计实体类、设计 DAO 层、设计 Service 层、设计 Controller 层、设计启动类、验证 JSON 数据和可视化页面。

针对需求一：统计分析客户端访问次数最多的前 10 个 IP 地址，按照访问次数降序排列，并采用 ECharts 柱状图可视化。

（1）搭建 Java Web 后台开发框架（SpringBoot）。

创建 SpringBoot 项目，操作过程如下。

① 打开 Eclipse，新建 Spring Boot → Spring Starter Project，单击"next"。

② 选择 jar 方式，设置 java 为"java8"，group 为"org.lanqiao"，Artifact 为"bigdata"，Package 为"org.lanqiao.bigdata"，单击"next"。

③ 选择事先要加载的组件：选择 SQL 里的"MyBatis Framework"和"MySQL Driver"，Web 里的"Spring Web"，依次单击"next"和"finish"。

④ 单击操作完成后，需要稍等一会儿，maven 会自动下载相关 jar 包资源。

(2)设计数据库连接。

打开 src/main/resources 目录下的 application.properties 文件,内容如程序清单 1.3 所示。

程序清单 1.3
```
spring.datasource.url=jdbc:mysql://localhost:3306/sqoop
spring.datasource.username=hive
spring.datasource.password=hive
spring.datasource.driver-class-name=com.mysql.cj.jdbc.Driver
```

(3)设计实体类。

创建包 org.lanqiao.bigdata.bean,然后在该包下创建类 A1,代码如程序清单 1.4 所示。

程序清单 1.4
```java
package org.lanqiao.bigdata.bean;

public class A1{
    private String name;
    private int num;

    public A1(String name,int num){
        super();
        this.name=name;
        this.num=num;
    }

    public String getName(){
        return name;
    }
    public void setName(String name){
        this.name=name;
    }
    public int getNum(){
        return num;
    }
    public void setNum(int num){
        this.num=num;
    }

    @Override
    public String toString() {
        return"A1[name="+name+",num="+num+"]";
    }
}
```

(4)设计 DAO 层。

创建包 org.lanqiao.bigdata.dao,然后在该包下创建接口 A1Mapper,代码如程序清单 1.5 所示。

程序清单 1.5
```java
package org.lanqiao.bigdata.dao;

import java.util.List;
import org.apache.ibatis.annotations.Mapper;
import org.apache.ibatis.annotations.Select;
import org.lanqiao.bigdata.bean.A1;
import org.springframework.stereotype.Repository;

@Mapper
@Repository
public interface A1Mapper {
    @Select("SELECT name,num FROM a1 order by num desc limit 10")
    List<A1> getA1();
}
```

(5)设计 Service 层。

创建包 org.lanqiao.bigdata.service,然后在该包下创建接口 A1Service,代码如程序清单 1.6 所示。

程序清单 1.6
```java
package org.lanqiao.bigdata.service;

import java.util.List;
import org.lanqiao.bigdata.bean.A1;

public interface A1Service {
    List<A1> getA1();
}
```

再创建实现类 A1ServiceImpl,代码如程序清单 1.7 所示。

程序清单 1.7
```java
package org.lanqiao.bigdata.service;

import java.util.List;
import org.lanqiao.bigdata.bean.A1;
```

```
import org.lanqiao.bigdata.dao.A1Mapper;
import org.springframework.beans.factory.annotation.Autowired;
import org.springframework.stereotype.Service;

@Service
public class A1ServiceImpl implements A1Service {
    @Autowired
    A1Mapper a1Mapper;

    @Override
    public List<A1> getA1() {
        return a1Mapper.getA1();
    }
}
```

（6）设计 Controller 层。

创建包 org.lanqiao.bigdata.controller，然后在该包下创建类 A1Controller，代码如程序清单 1.8 所示。

程序清单 1.8

```
package org.lanqiao.bigdata.controller;

import java.util.HashMap;
import java.util.List;
import java.util.Map;
import org.lanqiao.bigdata.bean.A1;
import org.lanqiao.bigdata.service.A1Service;
import org.springframework.beans.factory.annotation.Autowired;
import org.springframework.web.bind.annotation.RequestMapping;
import org.springframework.web.bind.annotation.RestController;

@RestController
public class A1Controller {
    @Autowired
    A1Service a1Service;

    @RequestMapping("/a1")
    public Map<String,Object> getA1(){
        List<A1> list=a1Service.getA1();
        String[] x=new String[list.size()];
        Integer[] y=new Integer[list.size()];
```

```
            for(int i=0;i<list.size();i++) {
                A1 in=list.get(i);
                x[i]=in.getName();
                y[i]=in.getNum();
            }
            Map map=new HashMap<String,Object>();
            map.put("x",x);
            map.put("y",y);
            return map;
    }
}
```

（7）设计启动类。

在包 org.lanqiao.bigdata 下创建启动类 DemoApplication，代码如程序清单 1.9 所示。

程序清单 1.9

```
package org.lanqiao.bigdata;

import org.springframework.boot.SpringApplication;
import org.springframework.boot.autoconfigure.SpringBootApplication;
import org.springframework.context.annotation.ComponentScan;

@SpringBootApplication
@ComponentScan(basePackages={"org.lanqiao.bigdata"})
public class DemoApplication {
    public static void main(String[] args) {
        SpringApplication.run(DemoApplication.class,args);
    }
}
```

（8）验证 JSON 数据。

右击项目 → 选择 "run as" → 选择 "spring boot app"，打开浏览器，输入地址 "http://localhost:8080/a1"，查看 JSON 数据是否如下所示。

```
{"x":["114.248.38.224","60.23.128.118","114.248.50.124","101.226.68.213",
"101.226.65.102","180.153.163.190","101.226.33.201","101.226.89.115","101.226.
89.121","101.226.99.197"],
"y":[283,74,26,3,2,2,1,1,1,1]}
```

（9）可视化页面。

下面按照需求用 ECharts 柱状图来呈现。

在项目 static 目录下创建 a1.html，代码如程序清单 1.10 所示。

程序清单 1.10

```html
<!DOCTYPE html>
<html>
<head>
<meta charset="utf-8">
<title>ECharts</title>
<!-- 引入 echarts.js -->
<script src="js/echarts.js"></script>
<!-- 引入 jquery-->
<script src="js/jquery-1.8.3.min.js"></script>
</head>
<body>
<!-- 为 ECharts 准备一个具备大小（宽高）的图表显示区域 -->
<div id="main" style="width:1400px;height:600px;"></div>
<script type="text/javascript">
        // 基于准备好的div 容器，初始化 ECharts 实例
        var myChart=echarts.init(document.getElementById('main'));

        // 指定图表的配置项和数据
        option={
            // 标题
            title:{
                text:'Top10 排名 '
            },
            // 图例
            legend:{
                data:[' 排名 ']
            },
            //x 轴内容
xAxis:{
     type:'category',
     axisLabel:{// 文字显示
             interval:0,// 间隔 0 表示每一个都强制显示
             rotate:40// 旋转角度 [-90,90]
     },
     data:[] // 数据由 jquery 请求后端服务器获取
},
            // 设置绘图网格，单个 grid 内最多可以放置上下两个 x 轴、左右两个 y 轴，设置绘图网格离左边和下边的距离
            // 这里主要用于 axisLabel 文字显示
```

```
            grid:{
                left:'10%',
                bottom:'35%'
            },
            //y轴内容
yAxis:{
type:'value',
            axisLabel:{
                    formatter:'{value}'
                }
},
            //提示框内容由坐标轴触发
            tooltip:{
                trigger:'axis',
                axisPointer:{
                    type:'shadow'
                }
            },
            //绘图类型和数据
            series:[{
                name:'排名',
                data:[],//数据由jquery请求后端服务器获取
                type:'bar',//柱状图
itemStyle:{
        normal:{
        color:function(params){
          var colorList=['red','orange','yellow','green','blue','purple','black','#123456','#789123','gray'];
            return colorList[params.dataIndex]
        }
        }
        }
            }]
        };

        //使用刚指定的配置项和数据显示图表
myChart.setOption(option);

        //使用jquery异步加载数据
        $.get('a1').done(function(data) {
myChart.setOption({//更新数据
xAxis:{
        type:'category',
        axisLabel:{//文字显示
```

```
                interval:0,// 间隔0表示每一个都强制显示
                rotate:40// 旋转角度[-90,90]
            },
            data:data.x //x轴数据
        },
        series:[{
            name:'排名',
            data:data.y,//y轴数据
            type:'bar',// 柱状图
        itemStyle:{
            normal:{
            color:function(params){
                var colorList=['red','orange','yellow','green','blue','purple','black',
'#123456','#789123','gray'];
                return colorList[params.dataIndex]
            }
            }
            }
        }]
        });
        });
    </script>
    </body>
    </html>
```

代码编写完毕，项目启动并运行后，打开浏览器，输入地址"http://localhost:8080/a1.html"，就能看到可视化数据了，如图1.8所示。

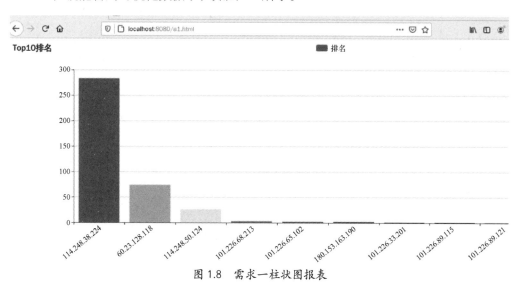

图1.8 需求一柱状图报表

针对需求二：分析客户端访问页面中不正常的页面（页面返回状态码不为 200），统计访问次数最多的前 10 个不正常页面的访问路径，按照访问次数降序排列，并采用 ECharts 折线图可视化。

（1）搭建 Java Web 后台开发框架（SpringBoot）。

项目后台之前已经搭建完毕。

（2）设计数据库连接。

数据库连接参数保持不变。

（3）设计实体类。

创建包 org.lanqiao.bigdata.bean，然后在该包下创建类 A2，代码如程序清单 1.11 所示。

程序清单 1.11

```java
package org.lanqiao.bigdata.bean;

public class A2 {
    private String name;
    private int num;

    public A2(String name,int num) {
        super();
        this.name=name;
        this.num=num;
    }

    public String getName() {
        return name;
    }
    public void setName(String name) {
        this.name=name;
    }
    public int getNum() {
        return num;
    }
    public void setNum(int num) {
        this.num=num;
    }
    @Override
    public String toString() {
        return "A2 [name="+name+",num="+num+"]";
    }
}
```

（4）设计 DAO 层。

创建包 org.lanqiao.bigdata.dao，然后在该包下创建接口 A2Mapper，代码如程序清单 1.12 所示。

程序清单 1.12

```
package org.lanqiao.bigdata.dao;

import java.util.List;
import org.apache.ibatis.annotations.Mapper;
import org.apache.ibatis.annotations.Select;
import org.lanqiao.bigdata.bean.A2;
import org.springframework.stereotype.Repository;

@Mapper
@Repository
public interface A2Mapper {
    @Select("SELECT name,num FROM a2 order by num desc limit 10")
    List<A2> getA2();
}
```

（5）设计 Service 层。

创建包 org.lanqiao.bigdata.service，然后在该包下创建接口 A2Service，代码如程序清单 1.13 所示。

程序清单 1.13

```
package org.lanqiao.bigdata.service;

import java.util.List;
import org.lanqiao.bigdata.bean.A2;

public interface A2Service {
    List<A2> getA2();
}
```

再创建实现类 A2ServiceImpl，代码如程序清单 1.14 所示。

程序清单 1.14

```
package org.lanqiao.bigdata.service;

import java.util.List;
import org.lanqiao.bigdata.bean.A2;
```

```
import org.lanqiao.bigdata.dao.A2Mapper;
import org.springframework.beans.factory.annotation.Autowired;
import org.springframework.stereotype.Service;

@Service
public class A2ServiceImpl implements A2Service {
    @Autowired
    A2Mapper a2Mapper;

    @Override
    public List<A2> getA2() {
        return a2Mapper.getA2();
    }
}
```

（6）设计 Controller 层。

创建包 org.lanqiao.bigdata.controller，然后在该包下创建类 A2Controller，代码如程序清单 1.15 所示。

程序清单 1.15

```
package org.lanqiao.bigdata.controller;

import java.util.HashMap;
import java.util.List;
import java.util.Map;
import org.lanqiao.bigdata.bean.A2;
import org.lanqiao.bigdata.service.A2Service;
import org.springframework.beans.factory.annotation.Autowired;
import org.springframework.web.bind.annotation.RequestMapping;
import org.springframework.web.bind.annotation.RestController;

@RestController
public class A2Controller {
    @Autowired
    A2Service a2Service;

    @RequestMapping("/a2")
    public Map<String,Object> getA2(){
        List<A2> list=a2Service.getA2();
        String[] x=new String[list.size()];
        Integer[] y=new Integer[list.size()];
```

```
            for(int i=0;i<list.size();i++) {
                    A2 in=list.get(i);
                    x[i]=in.getName();
                    y[i]=in.getNum();
            }
            Map map=new HashMap<String,Object>();
            map.put("x",x);
            map.put("y",y);
            return map;
    }
}
```

（7）设计启动类。

启动类保持不变。

（8）验证 JSON 数据。

右击项目 → 选择"run as" → 选择"spring boot app"，打开浏览器，输入地址"http://localhost:8080/a2"，查看 JSON 数据是否如下所示。

```
{"x":["/","/BlueVacation/Controller/SubShopHandler.ashx","/BlueVacation/
View/FINANCE/FinancialAccounts/Sheet.aspx","/BlueVacation/Controller/
SecondRecommendHandler.ashx","/uploadfile/RetailSales/","/BlueVacation/View/
RETAILSALESMGR/Payment/Default.aspx"],
 "y":[12,9,5,2,2,1]}
```

（9）可视化页面。

下面按照需求用 ECharts 折线图来呈现。

在项目 static 目录下创建 a2.html，代码如程序清单 1.16 所示。

程序清单 1.16
```
<!DOCTYPE html>
<html>
<head>
<meta charset="utf-8">
<title>ECharts</title>
<!-- 引入 echarts.js -->
<script src="js/echarts.js"></script>
<!-- 引入 jquery-->
<script src="js/jquery-1.8.3.min.js"></script>
</head>
<body>
<!-- 为 ECharts 准备一个具备大小（宽高）的图表显示区域 -->
```

```html
<div id="main" style="width:1400px;height:600px;"></div>
<script type="text/javascript">
```
```javascript
        // 基于准备好的div容器，初始化ECharts实例
        var myChart=echarts.init(document.getElementById('main'));

        // 指定图表的配置项和数据
        option={
            // 标题
            title:{
                text:'Top10排名'
            },
            // 图例
            legend:{
                data:['排名']
            },
            //x轴内容
xAxis:{
    type:'category',
    axisLabel:{// 文字显示
        interval:0,// 间隔0表示每一个都强制显示
        rotate:40// 旋转角度[-90,90]
    },
    data:[] // 数据由jquery请求后端服务器获取
},
            // 设置绘图网格，单个grid内最多可以放置上下两个x轴、左右两个y轴，设置绘图网格离左边和下边的距离
            // 这里主要用于axisLabel文字显示
            grid:{
                left:'10%',
                bottom:'35%'
            },
            //y轴内容
yAxis:{
type:'value',
            axisLabel:{
                    formatter:'{value}'
                }
},
            // 提示框内容由坐标轴触发
            tooltip:{
                trigger:'axis',
```

```
                axisPointer:{
                    type:'shadow'
                }
            },
            // 绘图类型和数据
            series:[{
                name:'排名',
                data:[],// 数据由 jquery 请求后端服务器获取
                type:'line',// 折线图
itemStyle:{
    normal:{
        color:function(params){
            var colorList=['red','orange','yellow','green','blue','purple','black','#123456','#789123','gray'];
            return colorList[params.dataIndex]
        }
    }
}
            }]
        };

        // 使用刚指定的配置项和数据显示图表。
myChart.setOption(option);

        // 使用 jquery 异步加载数据
        $.get('a2').done(function(data) {
myChart.setOption({// 更新数据
xAxis:{
    type:'category',
    axisLabel:{// 文字显示
        interval:0,// 间隔 0 表示每一个都强制显示
        rotate:40// 旋转角度 [-90,90]
    },
    data:data.x //x 轴数据
},
series:[ {
    name:'排名',
    data:data.y,//y 轴数据
    type:'line',// 折线图
itemStyle:{
    normal:{
```

```
color:function(params){
        var colorList=['red','orange','yellow','green','blue','purple',
'black','#123456','#789123','gray'];
        return colorList[params.dataIndex]
        }
        }
        }
        } ]
    });
    });
</script>
</body>
</html>
```

代码编写完毕，项目启动并运行后，打开浏览器，输入地址"http://localhost:8080/a2.html"，就能看到可视化数据了，如图1.9所示。

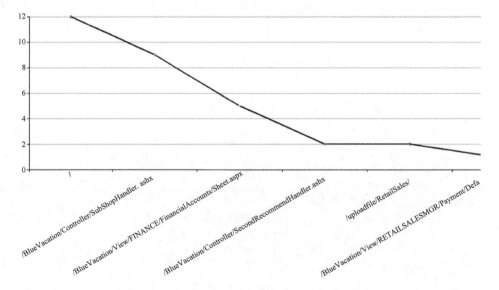

图1.9 需求二折线图报表

针对需求三：统计分析客户端访问次数最多的前10个访问路径，按照访问次数降序排列，并采用ECharts折线图可视化。

（1）搭建Java Web后台开发框架（SpringBoot）。

项目后台之前已经搭建完毕。

（2）设计数据库连接。

数据库连接参数保持不变。

（3）设计实体类。

创建包 org.lanqiao.bigdata.bean，然后在该包下创建类 A3，代码如程序清单 1.17 所示。

程序清单 1.17

```java
package org.lanqiao.bigdata.bean;

public class A3 {
    private String name;
    private int num;

    public A3(String name,int num) {
        super();
        this.name=name;
        this.num=num;
    }

    public String getName() {
        return name;
    }
    public void setName(String name) {
        this.name=name;
    }
    public int getNum() {
        return num;
    }
    public void setNum(int num) {
        this.num=num;
    }

    @Override
    public String toString() {
        return "A3[name="+name+",num="+num+"]";
    }
}
```

（4）设计 DAO 层。

创建包 org.lanqiao.bigdata.dao，然后在该包下创建接口 A3Mapper，代码如程序清单 1.18 所示。

程序清单 1.18

```
package org.lanqiao.bigdata.dao;

import java.util.List;
import org.apache.ibatis.annotations.Mapper;
import org.apache.ibatis.annotations.Select;
import org.lanqiao.bigdata.bean.A3;
import org.springframework.stereotype.Repository;

@Mapper
@Repository
public interface A3Mapper {
    @Select("SELECT name,num FROM a3 order by num desc limit 10")
    List<A3> getA3();
}
```

（5）设计 Service 层。

创建包 org.lanqiao.bigdata.service，然后在该包下创建接口 A3Service，代码如程序清单 1.19 所示。

程序清单 1.19

```
package org.lanqiao.bigdata.service;

import java.util.List;
import org.lanqiao.bigdata.bean.A3;

public interface A3Service {
    List<A3> getA3();
}
```

再创建实现类 A3ServiceImpl，代码如程序清单 1.20 所示。

程序清单 1.20

```
package org.lanqiao.bigdata.service;

import java.util.List;
import org.lanqiao.bigdata.bean.A3;
import org.lanqiao.bigdata.dao.A3Mapper;
import org.springframework.beans.factory.annotation.Autowired;
import org.springframework.stereotype.Service;
```

```
@Service
public class A3ServiceImpl implements A3Service {
    @Autowired
    A3Mapper a3Mapper;

    @Override
    public List<A3> getA3() {
        return a3Mapper.getA3();
    }
}
```

（6）设计 Controller 层。

创建包 org.lanqiao.bigdata.controller，然后在该包下创建类 A3Controller，代码如程序清单 1.21 所示。

程序清单 1.21

```
package org.lanqiao.bigdata.controller;

import java.util.HashMap;
import java.util.List;
import java.util.Map;
import org.lanqiao.bigdata.bean.A3;
import org.lanqiao.bigdata.service.A3Service;
import org.springframework.beans.factory.annotation.Autowired;
import org.springframework.web.bind.annotation.RequestMapping;
import org.springframework.web.bind.annotation.RestController;

@RestController
public class A3Controller {
    @Autowired
    A3Service a3Service;

    @RequestMapping("/a3")
    public Map<String,Object> getA3(){
        List<A3> list=a3Service.getA3();
        String[] x=new String[list.size()];
        Integer[] y=new Integer[list.size()];
        for(int i=0;i<list.size();i++) {
            A3 in=list.get(i);
            x[i]=in.getName();
            y[i]=in.getNum();
```

```
            }
            Map map=new HashMap<String,Object>();
            map.put("x",x);
            map.put("y",y);
            return map;
        }
    }
```

（7）设计启动类。

启动类保持不变。

（8）验证 JSON 数据。

右击项目 → 选择 "run as" → 选择 "spring boot app"，打开浏览器，输入地址 "http://localhost:8080/a3"，查看 JSON 数据是否如下所示。

```
{"x":["/BlueVacation/Controller/SupplierHandler.ashx","/BlueVacation/
Controller/RetailSalesHandler.ashx","/","/BlueVacation/Controller/
LanEmailHandler.ashx","/BlueVacation/Controller/NewsHandler.ashx","/sys/
ashx/ConfigHandler.ashx","/BlueVacation/Controller/SubShopHandler.ashx","/
BlueVacation/Controller/LvyouAreaHandler.ashx","/BlueVacation/Controller/
SubContractHandler.ashx","/BlueVacation/View/SUPPLIERMGR/Supplier/Search.html"],
"y":[30,25,24,24,24,24,22,22,16,15]}
```

（9）可视化页面。

下面按照需求用 ECharts 折线图来呈现。

在项目 static 目录下创建 a3.html，代码如程序清单 1.22 所示。

程序清单 1.22

```
<!DOCTYPE html>
<html>
<head>
<meta charset="utf-8">
<title>ECharts</title>
<!-- 引入 echarts.js -->
<script src="js/echarts.js"></script>
<!-- 引入 jquery-->
<script src="js/jquery-1.8.3.min.js"></script>
</head>
<body>
<!-- 为 ECharts 准备一个具备大小（宽高）的图表显示区域 -->
<div id="main" style="width:1400px;height:600px;"></div>
<script type="text/javascript">
```

```javascript
            // 基于准备好的div容器，初始化ECharts实例
            var myChart=echarts.init(document.getElementById('main'));

            // 指定图表的配置项和数据
            option={
                // 标题
                title:{
                    text:'Top10排名'
                },
                // 图例
                legend:{
                    data:['排名']
                },
                //x轴内容
xAxis:{
    type:'category',
    axisLabel:{// 文字显示
      interval:0,// 间隔0表示每一个都强制显示
      rotate:40// 旋转角度[-90,90]
    },
    data:[]  // 数据由jquery请求后端服务器获取
},
                // 设置绘图网格，单个grid内最多可以放置上下两个x轴、左右两个y轴，设置绘图网格离左边和下边的距离
                // 这里主要用于axisLabel文字显示
                grid:{
                    left:'10%',
                    bottom:'35%'
                },
                //y轴内容
yAxis:{
type:'value',
    axisLabel:{
                formatter:'{value}'
            }
},
                // 提示框内容由坐标轴触发
                tooltip:{
                    trigger:'axis',
                    axisPointer:{
                        type:'shadow'
```

```
                }
            },
            // 绘图类型和数据
            series:[{
                name:'排名',
                data:[],// 数据由jquery请求后端服务器获取
                type:'line',// 折线图
    itemStyle:{
        normal:{
        color:function(params){
                var colorList=['red','orange','yellow','green','blue','purple',
'black','#123456','#789123','gray'];
                return colorList[params.dataIndex]
        }
        }
        }
            }]
        };
            // 使用刚指定的配置项和数据显示图表。
    myChart.setOption(option);

            // 使用jquery异步加载数据
            $.get('a3').done(function(data) {
    myChart.setOption({// 更新数据
xAxis:{
        type:'category',
        axisLabel:{// 文字显示
            interval:0,// 间隔0表示每一个都强制显示
            rotate:40// 旋转角度[-90,90]
        },
        data:data.x  //x轴数据
},
series:[ {
        name:'排名',
        data:data.y,//y轴数据
        type:'line',// 折线图
    itemStyle:{
        normal:{
        color:function(params){
```

```
                var colorList=['red','orange','yellow','green','blue','purple','black',
'#123456','#789123','gray'];
                return colorList[params.dataIndex]
            }
        }
    }
    } ]
    });
    });
</script>
</body>
</html>
```

代码编写完毕，项目启动并运行后，打开浏览器，输入地址"http://localhost:8080/a3.html"，就能看到可视化数据了，如图1.10所示。

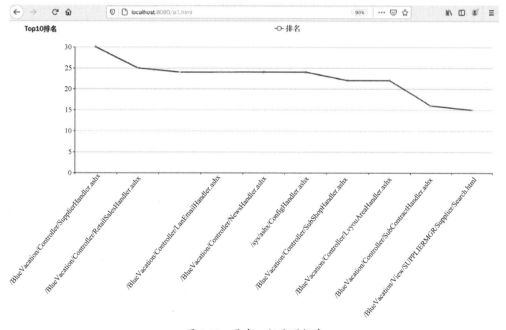

图1.10 需求三折线图报表

1.8 答疑解惑

问题1：有同学在运行项目 MapReduce 程序时，提示 "JobTracker is in safe mode"，该如何解决？

解答 1：JobTracker 处于安全模式下，只能进行读相关操作，不能进行写相关操作，需要退出安全模式，才能运行。一种方案是等待一段时间，系统会自动退出安全模式，另一种方案是使用如下命令退出安全模式。

```
shiyanlou:~/ $ hadoop dfsadmin -safemode leave
……
Safe mode is OFF
```

问题 2：请同学们思考 MapReduce 中 Combiner 的作用是什么？应用它的前提是什么？它和 Reducer 有什么区别？

解答 2：

（1）Combiner 的作用是对每一个 maptask 的输出进行局部汇总，以减少网络传输量。

（2）Combiner 能够应用的前提是不影响最终的业务逻辑，而且 Combiner 的输出 kv 类型应该与 Reducer 的输出 kv 类型对应。

（3）Combiner 和 Reducer 的区别在于运行的位置，Combiner 在每一个 maptask 所在的阶段运行，Reducer 接收全局所有 Mapper 的输出结果。

问题 3：运行 MapReduce 时，如果要用到第三方 jar 包该怎么办？

解答 3：可以使用以下几种方式。

（1）在 Hadoop 集群的每个节点上的 $HADOOP_HOME/lib 目录中加入 jar 文件（不推荐）。

（2）将第三方 jar 和程序打包到一个文件中，在集群中执行（不推荐）。

（3）通过命令行参数 -libjars 传递 jar 文件（推荐），例如：

```
shiyanlou:~/ $ hadoop jar PreProcess.jar org.lanqiao.bigdata.PreProcess
-libjars a.jar,b.jar/flume /out
```

（4）直接在 conf 中设置，如 conf.set（"tmpjars",*.jar），jar 文件用逗号隔开（推荐）。

问题 4：Eclipse 中运行 MapReduce 程序时，控制台无日志信息，该如何处理？

解答 4：复制 $HADOOP_HOME/etc/hadoop/ 目录下的 log4j.properties 文件到 MapReduce 项目 src 文件夹下。

问题 5：什么样的计算不能用 MapReduce 来提速？

解答 5：

（1）数据量很小。数据量小的话，可以直接单机计算，如果使用 MapReduce，反而显得笨重，计算时长会增加。

（2）复杂的小文件。小文件过多，不利于 MapReduce 计算，因为默认情况下一个小文件也会分配一个 mapTask，导致计算时长大幅增加，复杂的文件也会加大 MapReduce 的编写难度。

（3）事务处理。MapReduce 编程不涉及事务处理，没有回滚等机制。

（4）只有一台机器的时候。只有一台机器没必要使用 MapReduce 来计算。

问题 6：什么情况下使用 MapReduce，而不使用 Hive？

解答 6：一般来说，结构复杂、业务复杂的数据建议使用 MapReduce，不使用 Hive。

1.9 拓展练习

1．拓展项目

我们拓展一下本章项目，假设运维部门想知道访问哪些页面非常耗时，想进一步优化这些页面，于是我们给出如下需求：统计分析客户端访问页面所需平均时间最长的前 10 个访问路径，按照访问所需平均时间降序排列，并采用 ECharts 柱状图可视化。要求按照本章所讲的步骤来做，请同学们自行完成该任务。

2．强化 MapReduce 编程

使用 MapReduce 实现倒排索引（Inverted Index）。

（1）问题概述。

倒排索引是文档检索系统中最常用的数据结构，被广泛地应用于全文搜索引擎。它主要用来存储某个单词（或词组）在一个文档或一组文档中的存储位置的映射，即提供了一种根据内容来查找文档的方式。由于它不是根据文档来确定文档所包含的内容，而是进行相反的操作，因而被称为倒排索引。

通常情况下，倒排索引由一个单词（或词组）以及相关的文档列表组成，文档列表中的文档要么是标识文档的 ID，要么是文档所在位置的 URL，如下所示。

```
hello           a.txt,b.txt,c.txt
world           a.txt,d.txt
```

可以看出，单词"hello"出现在 a.txt 文档、b.txt 文档、c.txt 文档中，单词"world"出现在 a.txt 文档、d.txt 文档中。在实际应用中，还需要给每个文档添加一个权值，用来指出每个文档与搜索内容的相关度，最常见的是使用词频作为权值，即记录单词在文档中出现的次数，如下所示。

```
hello          a.txt:2,b.txt:3,c.txt:1
world          a.txt:3,d.txt:4
```

可以看出"hello"这个单词在 a.txt 文档中出现过 2 次,在 b.txt 文档中出现过 3 次,在 c.txt 文档中出现过 1 次。"world"这个单词在 a.txt 文档中出现过 3 次,在 d.txt 文档中出现过 4 次。

(2)编程实现。

假设目前有 2 个文件:a.txt 和 b.txt,内容如下。

```
shiyanlou:~/ $ cat a.txt
hello world
shiyanlou:~/ $ cat b.txt
hello world hello
```

请同学们编写 MapReduce 程序实现如下输出效果。

```
shiyanlou:~/ $ hadoop fs -cat /out/part*
hello          a.txt:1; b.txt:2
world          a.txt:1; b.txt:1
```

第 2 章 招聘网站数据分析项目

项目标签：职业院校技能大赛

【本章简介】

本章通过介绍高职院校职业技能大赛中的大数据技术与应用赛项，帮助读者熟悉这一赛项，掌握大数据相关的职业技能，提高大数据的处理能力。

【知识地图】

本章项目涉及的知识地图如图 2.1 所示，读者在学习过程中可以查询相关知识卡片，补充理论知识。

图 2.1 本章知识地图

2.1 大赛简介

1．竞赛目的

贯彻落实国务院发布的《促进大数据发展行动纲要》及工业和信息化部发布的《大数据产业发展规划（2016—2020年）》，加快实施国家大数据战略，推动大数据产业健康快速发展，针对高职"大数据技术与应用"专业建设和发展的需求，通过引入大数据各个环节的实际应用场景，全面考察大数据技术基础、软件开发相关技术、Hadoop及其生态组件部署与管理、数据采集、数据清洗、数据分析和数据可视化等前沿的知识、技术技能以及职业素养和团队协作能力。

赛项围绕大数据产业各个岗位的实际需求和要求进行设计，通过大赛搭建校企合作的平台，深化产教融合，提升大数据技术与应用专业及其他相关专业毕业生的能力，推动院校和企业联合培养大数据人才，加强学校教育与产业发展的有效衔接，促进职业院校信息类相关专业共同发展，为国家战略规划提供大数据领域的高素质技能型人才。

2．竞赛方式

比赛以团队方式进行，每个参赛队由1名领队（可由指导教师兼任）、2名指导教师、3名选手（其中队长1名）组成，指导教师须为本校专职教师，竞赛时长为4小时。

3．竞赛内容

赛项以大数据技术与应用为核心内容，重点考察参赛选手基于Hadoop平台环境，利用Hadoop技术生态组件，综合软件开发相关技术，解决实际问题的能力，概括来说包括以下内容。

（1）掌握按照项目需求配置、管理Hadoop大数据平台及相关生态组件的能力。

（2）掌握企业常用采集工具和网络"爬虫"的相关技术，完成指定数据采集及处理的能力。

（3）综合利用MapReduce、Spark等技术，以及分布式存储系统、数据仓库Hive等工具，使用Java、Python、Scala等开发语言，完成数据清洗、存储、转化、分析及数据推送等一系列大数据操作。

（4）综合运用HTML、CSS、JavaScript、Python等开发语言，对数据进行可视化呈现。

（5）根据数据可视化结果，完成数据分析报告的编写。

详细的竞赛内容如表2.1所示。

表 2.1　竞赛内容详解

考核环节	考核知识点和技能点	描述
Hadoop 平台及组件的部署管理	Hadoop 平台安装部署和基本配置	考核 Hadoop 平台及组件的部署能力，掌握常用的基本配置和命令，能够部署和管理 Hadoop 高可用集群
	Hadoop 集群节点的动态增加与删除	
	Hadoop 平台相关组件部署与管理	
	Hadoop 平台的高可用	
数据采集	使用开发者工具查看网页源码，分析网页结构，明确数据采集对象	考核多维度数据采集能力，包括对关系型数据库、非关系型数据库和网络"爬虫"技术的应用
	构建数据采集请求，抓取网络数据	
	利用网络"爬虫"相关组件实现网络数据"爬取"	
	规则文件数据和关系型数据库数据抓取以及数据同步	
	非关系型数据库数据抓取以及数据同步	
	数据采集结果导出及数据库推送	
数据清洗与分析	基于 Hadoop 平台架构组件和多维度的数据采集，实现数据一致性检查、无效值和缺省值的处理	考核对分布式计算、分布式存储系统、数据仓库等进行综合应用的能力，使用 Java、Python、Scala 等开发语言，完成数据清洗、数据存储、数据转化、数据分析、数据预测及数据推送等一系列数据操作
	多表数据合并和离群值处理	
	通过常见的数据分析算法，对数据进行标准化、离散化和多元化分析	
	掌握数据仓库的导入、导出，利用数据仓库相关命令或代码实现数据多维度、多层次的分析	
	对数据的查询、整理和计算。进行编译、打包、发布，执行程序，完成数据处理、清洗	
	实现不同数据库间的文件传输及转换	
	数据预测分析	
数据可视化	编写后台代码，实现数据库访问和数据整理	使用 Python 及 Web 前端等编程语言，通过常见的数据可视化方法，将数据分析结果以图表的形式进行呈现
	编写 Web 前端代码，对数据分析结果进行呈现	
综合分析	通过知识技能，根据数据分析、预测及可视化结果进行分析，做出分析报告	考核大数据技术与分析的综合操作能力和业务分析能力

注意：每次竞赛题目都涉及这五大考核环节，但是每一个环节中的考核知识点和技能点可能不一样。

4. 竞赛技术平台

（1）硬件设备。

竞赛硬件设备如表 2.2 所示。

表 2.2　竞赛硬件设备

序号	设备名称	数量	备注
1	服务器	1	支撑大数据竞赛管理系统运行，内嵌虚拟化资源管理控制端，作为虚拟化资源管理系统的计算资源、网络资源和存储资源的源节点。 ① CPU 模块：2×2.3GHz。 ② 内存模块：8×32GB。 ③ 硬盘模块：6×600GB SAS 10K。 ④ 网口：4 端口千兆电接口网卡 -360T-B2。 ⑤ 1+1 冗余电源
2	大数据竞赛平台	1	系统基于 kvm 构建，可模拟大数据环境搭建、大数据采集、大数据预处理、大数据存储及管理、大数据分析及挖掘、大数据展现和应用等贯穿大数据技术的相关知识点，提供大数据竞赛管理系统所需的虚拟服务器，结构化、半结构化及非结构化数据的数据库等基础支撑环境；涵盖分布式虚拟存储技术，大数据获取、存储、组织、分析和决策操作的可视化技术。具体包括 Hadoop、HDFS、Hbase、Hive、MapReduce、Kafka、Spark、Storm、Mahout、MySQL、ECharts 等。所涉及开发语言包括 Java、Python、Scala、HTML、JavaScript 等
3	PC	3	竞赛选手比赛使用。性能相当于 i5 处理器，8GB 以上内存，1TB 以上硬盘，显示器分辨率要求在 1024 像素×768 像素以上
4	交换机	1	① 机架式交换机。 ② 端口：24 个 10/100/1000Base-TX 以太网端口。 ③ 速度：10/100/1000Base。 ④ 全千兆三层交换机，支持访问控制

（2）软件环境。

竞赛设备安装的软件如表 2.3 所示。

表 2.3　竞赛软件环境

设备类型	软件类别	软件名称、版本号
服务器集群	大数据集群操作系统	CentOS 7.4
	大数据分析平台组件	Hadoop 2.6.0
		Yarn 2.6.0
		ZooKeeper 3.4.5
		Hive 1.1.0
		Flume 1.6.0
		Sqoop 1.4
		Kafka 1.0
		Spark 2.0
	数据库	MySQL 5.7
开发客户端	PC 操作系统	Windows 10（64 位）
	浏览器	Chrome
	开发语言	Python 3.6（64 位）
		Java 8
		Scala 12
	开发工具	Pycharm 2019（Community Edition）
		IDEA 2019（Community Edition）
	数据采集组件	Requests
		Scrapy
	数据可视化组件	ECharts 4.0.4
		Flask
		Jinja2
		Matplotlib
	Xshell	免费版
	Xftp	免费版
	文档编辑器	Office 2007 以上
	输入法	拼音输入法

2.2 任务一：需求分析

1．业务背景

近年来，随着 IT 产业的加速发展，全国各地对 IT 类人才的需求也越来越多，某公司为了明确今后 IT 产业人才的培养方向，在多地进行 IT 公司岗位情况调研分析。你所在的小组将承担模拟调研分析的任务，通过在招聘网站进行招聘信息的"爬取"，获取公司名称、工作地点、岗位名称、招聘要求、招聘人数等信息，并通过对数据的清洗和分析，得出各城市大数据相关职位招聘人数、数量差异及平均薪资情况，并绘制相关图表展示。

2．功能需求

（1）Hadoop 相关组件安装部署。

当前已搭建 Hadoop 运行环境，并安装了 MySQL 数据库，请在此基础上按照相关操作步骤安装 Hive 组件。

① 将指定路径下的 Hive 安装包解压缩并重命名。

② 设置 Hive 环境变量。

③ 编辑 Hive 相关配置文件。

④ 初始化 Hive 元数据。

⑤ 启动 Hive。

（2）数据采集。

① 从指定招聘网站中抓取数据，提取有效数据项，并保存为 JSON 格式文件。

② 设置 get 请求参数并将信息返回给变量 response。

③ 将提取数据转化成 JSON 格式，并赋值变量。

④ 用 with 函数创建 JSON 文件，通过 JSON 方法，写入 JSON 数据。

⑤ "爬取"的数据需要导入 Hadoop 平台进行数据清洗与分析，在 HDFS 文件系统中创建文件夹，并将 JSON 文件上传到该文件夹下。

（3）数据清洗与分析。

① 为便于数据分析与可视化，需要对"爬取"的数据进行清洗，使用 Java 语言编写数据清洗的 MapReduce 程序。

② 将清洗程序上传至 Hadoop，并对 HDFS 的原始数据进行清洗。

③ 将清洗后的数据加载到 Hive 数据仓库中。

④ 通过运行 HQL 命令完成数据分析统计。

（4）数据导出到 MySQL。

将数据分析结果导出到 MySQL 表中。

（5）数据可视化。

为更好地将数据分析结果呈现出来，需要对数据分析的结果进行可视化呈现，本次数据可视化需要呈现 3 个部分的内容。

① 按要求使用柱状图展示各城市大数据相关职位招聘人数，并在前端显示。要求显示结果包含以下内容。

主标题：各城市大数据相关职位招聘人数。

副标题：招聘人数变化趋势。

横坐标：城市信息。

纵坐标：招聘人数。

② 按要求使用折线图展示大数据各相关职位招聘数量，并在前端显示。要求显示结果包含以下内容。

主标题：大数据各相关职位招聘数量。

副标题：招聘数量变化趋势。

横坐标：职位名称。

纵坐标：职位数量。

③ 按要求使用柱状图展示大数据相关职位平均薪资，并在前端显示。要求显示结果包含以下内容。

主标题：大数据相关职位平均薪资。

副标题：职位薪资差别。

横坐标：职位名称。

纵坐标：平均薪资（万元／月）。

（6）完成分析报告。

请结合数据分析结果回答以下问题：

① 根据分析结果，大数据岗位需要哪些主要技能？为什么？

② 根据分析结果，各地大数据产业发展情况如何？

③ 根据市场需求分析，大数据行业有哪些人才培养方向？为什么？

④ 今后大数据产业地域发展方向是什么？

2.3 任务二：项目流程

1．技术领域判定

从需求分析中，我们可以了解到该项目有以下几个特征。

（1）数据量较大。

（2）不需要实时分析。

（3）需要有简单的报表。

本章的项目是一个从网络"爬取"数据后进行统计分析的场景，解决方案是典型的大数据离线计算技术。

2．技术架构分析

该项目的大数据架构由下往上可以分为 4 层：数据采集层、数据存储层、数据分析层、数据展示层。下面我们对这 4 层做一下简单分析。

（1）数据采集层。我们需要的原始数据来源于某个大型招聘网站。通过"爬取"的方式来获取数据，然后存储到数据存储层中。这里我们使用 Python 中的 requests 模块来采集数据。

（2）数据存储层。使用目前应用最广泛的大数据分析平台 Hadoop 的文件系统 HDFS 来存储数据。

（3）数据分析层。该层技术较多，这里我们使用 MapReduce 来做数据清洗（预处理）工作，使用 Hive 来做数据分析工作。

（4）数据展示层。展示层技术较多，包括 ECharts、Matplotlib 等，这里我们使用简单易用又非常强大的 ECharts，配合 Python Web 框架 Flask，以图形化的方式来展示数据结果，如饼图、柱状图、折线图、雷达图等。

3．实现流程

本项目的流程，如图 2.2 所示。先使用 Python "爬取"网页数据到 HDFS，然后使用 MapReduce 进行数据清洗，清洗后的数据存入 HDFS，然后使用 Hive 进行数据分析，分析结果存入 HDFS，之后使用 Sqoop 将数据分析结果导出到 MySQL 中，再创建 Flask Web 应用程序来访问 MySQL 数据库，并返回 JSON 数据，最后使用 ECharts 呈现在网页上。

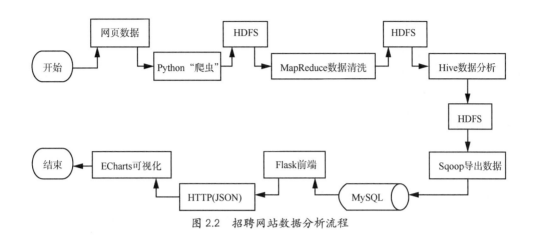

图 2.2　招聘网站数据分析流程

4．开发环境准备

在完成后续任务之前,我们需要安装并配置一系列的软件,准备好开发环境,如表 2.4 所示。这些基础操作,我们放入附录中,以供不熟悉的同学参考。

表 2.4　开发环境准备

操作	附录
Hadoop 安装部署和配置	附录 1
Hive 安装部署和配置	附录 6
Sqoop 安装部署和配置	附录 7
Hadoop 高可用集群环境安装部署和配置	附录 8
Hadoop 集群节点动态管理	附录 9

2.4　任务三：使用 Python "爬取" 招聘网站数据

使用 Python "爬取" 招聘网站数据的流程如图 2.3 所示。

图 2.3　使用 Python "爬取" 招聘网站数据的流程

按照操作步骤,可以将 Python "爬取" 招聘网站数据的过程划分为 3 个阶段,分别是分析网站源码结构、编写 Python "爬虫" 和数据写入 JSON 文件。

1. 分析网站源码结构

根据需求，我们使用 Python 中的 requests 模块来"爬取"网站数据。网站由竞赛组委会提供，是专门用来"爬取"数据的模拟网站，类型为招聘网站，部分界面如图 2.4 所示。

我们"爬取"的数据限定在大数据相关职位内，考虑到我们的 3 个需求，只"爬取"有用的相关信息：地区、职位、薪资、人数、日期。实际竞赛中数据量非常大，图中展示数据有限。竞赛数据并没有限制职位所在的城市，我们这里为了演示方便，只选取 8 个有代表性的城市：北京、上海、广州、深圳、武汉、杭州、成都、西安。职位选取那些职能明确、彼此能区分开的职位：大数据开发工程师、大数据分析工程师、大数据算法工程师、大数据架构师、大数据项目经理、大数据运维工程师、大数据其他相关职位。薪资通常为一个范围，如 1.2 万～1.8 万元/月，可以取中值 1.5 万元/月。人数为该岗位的实际招聘数量，如招 5 人。日期为几月几日，如 3 月 25 日发布。

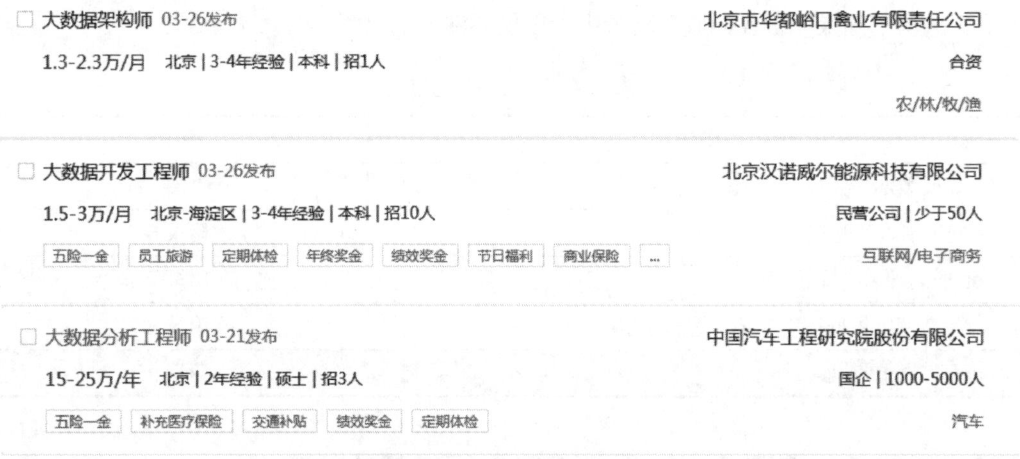

图 2.4　招聘网站部分界面

图 2.4 对应的源代码片段如下所示。

```
<div class="list">
    <a href="https:// 某 网 址 /beijing/138902499.html?jobtype=0_0&rc=01" class="e e3">
        <b class="jobid" value="138902499" jobtype="0_0"></b>
        <i>1.3-2.3万元 / 月 </i>
        <strong><span> 大数据架构师 </span></strong>
        <em>03-26 发布 </em>
```

```html
        <p>
            <span>北京</span>|
            <span>3-4年经验</span>|
            <span>本科</span>|
            <span>招1人</span>
        </p>
        <button class="btnbtn_jobid" jobtype="0_0" value="138902499" onclick="clicktoapply(this);return false;">申请</button>
        <aside>北京市华都峪口禽业有限责任公司</aside>
    </a>
    <a href="https://某网址/beijing/138902500.html?jobtype=0_0&rc=02" class="e e3">
        <b class="jobid" value="138902500" jobtype="0_0"></b>
        <i>1.5-3万元/月</i>
        <strong><span>大数据开发工程师</span></strong>
        <em>03-26发布</em>
        <p>
            <span>北京-海淀区</span>|
            <span>3-4年经验</span>|
            <span>本科</span>|
            <span>招10人</span>
        </p>
        <button class="btnbtn_jobid" jobtype="0_0" value="138902500" onclick="clicktoapply(this);return false;">申请</button>
        <aside>北京汉诺威尔能源科技有限公司</aside>
    </a>
    <a href="https://某网址/beijing/138902501.html?jobtype=0_0&rc=03" class="e e3">
        <b class="jobid" value="138902501" jobtype="0_0"></b>
        <i>15-25万元/年</i>
        <strong><span>大数据分析工程师</span></strong>
        <em>03-21发布</em>
        <p>
            <span>北京</span>|
            <span>2年经验</span>|
            <span>硕士</span>|
            <span>招3人</span>
        </p>
        <button class="btnbtn_jobid" jobtype="0_0" value="138902501" onclick="clicktoapply(this);return false;">申请</button>
        <aside>中国汽车工程研究院股份有限公司</aside>
    </a>
    ……
</div>
```

2. 编写 Python "爬虫"

我们使用开发工具 PyCharm 来开发网络"爬虫",创建一个项目,然后创建一个文件 main.py,其内容如程序清单 2.1 所示。

程序清单 2.1

```python
import requests
from lxml import etree
import json
import time
import random

citys=[]  #所有城市
titles=[]  #所有职位
salarys=[]  #所有薪水
nums=[]  #所有人数
times=[]  #所有时间

def parseUrl(url):
    headers={
        'User-Agent':'Opera/9.80 (Android 2.3.4; Linux; Opera Mobi/ADR-1301071546) Presto/2.11.355 Version/12.10'
    }
    res=requests.get(url,headers=headers)
    res.encoding='utf-8'
    selector=etree.HTML(res.text)

    city=selector.xpath('//*[@class="e e3"]/p/span[1]/text()')  #北京
    title=selector.xpath('//*[@class="e e3"]/strong/span/text()')  #大数据开发工程师
    salary=selector.xpath('//*[@class="e e3"]/i/text()')  #1.5-3万元/月
    num=selector.xpath('//*[@class="e e3"]/p/span[last()]/text()')  #招10人
    time=selector.xpath('//*[@class="e e3"]/em/text()')  #03-25 发布

    #每一页内容追加到总的列表中
    citys.extend(city)
    titles.extend(title)
    salarys.extend(salary)
    nums.extend(num)
    times.extend(time)
```

```python
if __name__ == '__main__':
    city={'010000':'北京','020000':'上海','030200':'广州','040000':'深圳',
          '180200':'武汉','080200':'杭州','090200':'成都','200200':'西安'}
    cityKeys=list(city.keys())
    key='大数据'
    #每个城市取前 50 页 URL
    for i in range(len(cityKeys)):
        for j in range(1,51):
            parseUrl('https://某网址:8080/list/' + cityKeys[i] + ',000000,0000,00,9,99,' + key + ',2,' + str(j) +'.html?lang=c&postchannel=0000&workyear=99&cotype=99&degreefrom=99&jobterm=99&companysize=99&ord_field=0&dibiaoid=0&line=&welfare=')
            #设置线程休眠时间，防止被"反爬虫"
            time.sleep(random.randrange(1,4))
    #写入 JSON 文件
    data=[]
    t={}
    for i in range(len(citys)):
        t["city"]=citys[i]
        t["title"]=titles[i]
        t["salary"]=salarys[i]
        t["num"]=nums[i]
        t["time"]=times[i]
        data.append(t)
    with open("data.json","w") as f:
        json.dump(data,f)
```

我们来解释一下其中的核心代码。

（1）网站中需要爬取的网页地址 URL：'https://某网址/list/' + cityKeys[i] + ',000000,0000,00,9,99,' + key + ',2,' + str(j) + '.html？lang=c&postchannel=0000&workyear=99&cotype=99&degreefrom=99&jobterm=99&companysize=99&ord_field=0&dibiaoid=0&line=&welfare='，其中 cityKeys[i] 为城市名，str(j) 为第几页，通过 2 层循环遍历 8 个城市前 50 页数据。

（2）每一页有 100 条招聘信息，使用 XPath 技术来"爬取"需要的信息，包括城市、职位、薪资、招聘人数、发布时间。比如，title=selector.xpath('//*[@class="e e3"]/strong/span/text( )')，"爬取"的就是每个页面中的所有招聘职位，返回值为一个数组，包含该页面中所有的 100 个职位。

（3）每"爬取"一个页面的信息后，就把信息存储到一个全局的数组中，如 titles.

extend(title)。

（4）全部"爬取"完毕后，遍历全局数组，把每条招聘信息封装成一个字典对象，如 {"city":" 北京 ","title":" 大数据开发工程师 ","salary":"1.5-3 万元 / 月 ","num":" 招 1 人 ","time":"03-26 发布 "}，然后再把所有的招聘信息封装成一个 JSON 数组，如 [{…},{…},…]，写进 JSON 文件中。

3．数据写入 JSON 文件

采集完毕后，数据被写入 data.json 文件中，格式如下。

```
[hadoop@server1 ~]$ more data.json
[{"city":"北京","title":"大数据架构师","salary":"1.3-2.3万元/月","num":"招1人",
"time":"03-26发布"},{"city":"北京","title":"大数据开发工程师","salary":"1.5-2.5万
元/月","num":"招5人","time":"03-26发布"},{"city":"北京","title":"大数据算法工程
师","salary":"1.2-2万元/月","num":"招1人","time":"03-26发布"},{"city":"北京",
"title":"大数据开发工程师","salary":"1.5-2万元/月","num":"招1人","time":"03-26发布
"},{"city":"北京","title":"大数据开发工程师","salary":"1.5-2万元/月","num":"招10人",
"time":"03-26发布"},{"city":"北京","title":"大数据开发工程师","salary":"2.5-5万元/
月","num":"招2人","time":"03-26发布"},{"city":"北京","title":"大数据开发工程师",
"salary":"24-32万元/年","num":"招2人","time":"03-26发布"},......]
```

然后上传到 HDFS 上的目录 /input 中。

```
[hadoop@server1 ~]$ hadoop fs -mkdir /input
[hadoop@server1 ~]$ hadoop fs -put data.json /input
```

2.5 任务四：使用 MapReduce 预处理数据

本任务涉及 HDFS Shell 操作和 MapReduce 编程，对此不熟悉的同学可以参考表 2.5 中提到的附录进行学习。

表 2.5　HDFS Shell 操作和 MapReduce 编程

操作	附录
掌握 HDFS Shell 操作	附录 2
通过 WordCount 熟悉 MapReduce	附录 3
深入理解 MapReduce	附录 4

使用 MapReduce 清洗数据的流程如图 2.5 所示。

图 2.5 使用 MapReduce 清洗数据的流程

按照操作步骤,可以将 MapReduce 清洗数据的过程划分为 4 个阶段,分别是分析数据问题、编写 MapReduce 程序、打包执行和验证数据结果。

1. 分析数据问题

项目中采集的数据可能存在数据缺失、数据格式不良好、数据不正确、数据前后不一致等数据质量问题,会影响我们后续的数据分析操作,甚至导致错误。因此我们需要先对原始数据进行一次有效清洗。

通过观察采集到的原始数据,发现主要有以下几个问题。

(1)城市有"北京""北京—海淀区",应统一设置为"北京"。

(2)职位由于是不同公司自己设定的,有很多细小的差异,需要灵活处理。比如含有"开发"或"研发"的统一设置为"大数据开发工程师",含有"分析"的统一设置为"大数据分析工程师",含有"算法"的统一设置为"大数据算法工程师",含有"架构"的统一设置为"大数据架构师",含有"经理"的统一设置为"大数据项目经理",含有"运维"的统一设置为"大数据运维工程师",其余的统统归于"大数据其他相关职位"。

(3)招聘人数为"招 3 人",需要把"招"和"人"去掉,只保留数字。若"招若干人",直接放弃整条数据。

(4)薪资有 1.5 万~ 2 万元 / 月和 24 万~ 32 万元 / 年,统一换算成月薪,薪资范围换成中值。

(5)发布时间有早有晚,只取最近 1 周发布的招聘信息,最近的数据是"3 月 26 日发布",那么最早的数据定为"3 月 20 日发布"的数据,之前的数据就不用了。这个需求不是大赛题目中的需求,是笔者补充的,因为太早的数据不具有代表性和时效性,实际大赛中一定要按照题目要求来设定。

2. 编写 MapReduce 程序

(1)创建 maven 项目。

打开 Eclipse,新建一个 maven project,操作步骤如下。

右击项目浏览器 → 新建项目 → 选择"maven project"。设置 catalog 为"internal",项目类型为"maven-archetype-quickstart",Group Id 为"org.lanqiao",Artifact Id 为"bigdata",

单击"finish"。

为 pom.xml 配置本项目对 jar 包的依赖：

```xml
<dependencies>
    <dependency>
        <groupId>junit</groupId>
        <artifactId>junit</artifactId>
        <version>4.12</version>
    </dependency>
    <dependency>
        <groupId>log4j</groupId>
        <artifactId>log4j</artifactId>
        <version>1.2.17</version>
    </dependency>
    <dependency>
        <groupId>org.apache.hadoop</groupId>
        <artifactId>hadoop-client</artifactId>
        <version>2.9.2</version>
    </dependency>
    <dependency>
        <groupId>org.apache.hadoop</groupId>
        <artifactId>hadoop-common</artifactId>
        <version>2.9.2</version>
    </dependency>
    <dependency>
        <groupId>com.alibaba</groupId>
        <artifactId>fastjson</artifactId>
        <version>1.2.73</version>
    </dependency>
</dependencies>
```

其中 fastjson 是用来处理 JSON 数据的包，由于竞赛中没有硬性规定，可以自己选取熟悉的技术，如阿里巴巴的 FastJson、谷歌的 Gson 等。

（2）编写 MapReduce 程序。

新建类 PreProcess，完整的代码如程序清单 2.2 所示。

程序清单 2.2

```java
package org.lanqiao.bigdata;

import java.net.URI;
import java.text.NumberFormat;
```

```java
import java.text.ParseException;
import java.text.SimpleDateFormat;
import java.util.Date;

import org.apache.hadoop.conf.Configuration;
import org.apache.hadoop.fs.FileSystem;
import org.apache.hadoop.fs.Path;
import org.apache.hadoop.io.LongWritable;
import org.apache.hadoop.io.NullWritable;
import org.apache.hadoop.io.Text;
import org.apache.hadoop.mapreduce.Job;
import org.apache.hadoop.mapreduce.Mapper;
import org.apache.hadoop.mapreduce.lib.input.FileInputFormat;
import org.apache.hadoop.mapreduce.lib.output.FileOutputFormat;

import com.alibaba.fastjson.JSON;
import com.alibaba.fastjson.JSONArray;
import com.alibaba.fastjson.JSONObject;

public class PreProcess {
    public static void main(String[] args) throws Exception {
        Configuration conf=new Configuration();
        // 获取集群访问路径
        FileSystemfileSystem=FileSystem.get(new URI("hdfs://localhost:9000/"),conf);
        // 判断输出路径是否已经存在，如果存在则删除
        Path outPath=new Path(args[1]);
        if(fileSystem.exists(outPath)){
            fileSystem.delete(outPath,true);
        }
        // 生成job，并指定job的名称
        Job job=Job.getInstance(conf,"PreProcess");
        // 指定打成jar后的运行类
        job.setJarByClass(PreProcess.class);
        // 指定mapper类
        job.setMapperClass(MyMapper.class);
        // 指定mapper的输出类型
        job.setMapOutputKeyClass(Text.class);
        job.setMapOutputValueClass(NullWritable.class);
```

```java
            //数据的输入目录
            FileInputFormat.addInputPath(job,new Path(args[0]));
            //数据的输出目录
            FileOutputFormat.setOutputPath(job,new Path(args[1]));
            System.exit(job.waitForCompletion(true)?0:1);
    }

    static class MyMapper extends Mapper<LongWritable,Text,Text,NullWritable>{
            protected void map(LongWritable k1,Text v1,Context context) throws java.io.IOException ,InterruptedException {
                    //使用fastjson把原始数据解析为JSON数组
                    JSONArray arr=JSON.parseArray(v1.toString());
                    String num="";//人数
                    String time="";//时间
                    String city="";//城市
                    String title="";//职位
                    String salary="";//薪资
                    SimpleDateFormat sdf=new SimpleDateFormat("yyyy-MM-dd");
                    NumberFormat nf=NumberFormat.getNumberInstance();
                    nf.setMaximumFractionDigits(2);//保留2位小数
                    for(int i=0;i<arr.size();i++) {
                            JSONObject o=arr.getJSONObject(i);
                            //处理人数
                            num=o.getString("num");
                            if(num.contains("招若干人")) {
                                    continue;
                            }else {
                                    num=num.substring(1,num.length()-1);
                            }
                            //处理时间
                            time=o.getString("time");
                            time="2022-"+time.substring(0,time.length()-2);
                            try {
                                    Date d=sdf.parse(time);
                                    if(d.before(new Date(122,2,20))) {//2022-03-20
                                            continue;
                                    }
                            } catch (ParseException e) {
                                    e.printStackTrace();
```

```java
            }
            // 处理城市
            city=o.getString("city");
            int index=city.indexOf("-");
            if(index>=0) {
                    city=city.substring(0,index);
            }
            // 处理职位
            title=o.getString("title");
            if(title.contains("开发")||title.contains("研发")) {
                    title=" 大数据开发工程师 ";
            }else if(title.contains("分析")) {
                    title=" 大数据分析工程师 ";
            }else if(title.contains("算法")) {
                    title=" 大数据算法工程师 ";
            }else if(title.contains("架构")) {
                    title=" 大数据架构师 ";
            }else if(title.contains("经理")) {
                    title=" 大数据项目经理 ";
            }else if(title.contains("运维")) {
                    title=" 大数据运维工程师 ";
            }else {
                    title=" 大数据其他相关职位 ";
            }
            // 处理薪资
            salary=o.getString("salary");
            String[] sa=salary.split("-|万元/");
            if(sa[2].equals("月")) {
                    salary=nf.format((Float.parseFloat(sa[0])+Float.parseFloat(sa[1]))/2)+"";
            }else if(sa[2].equals("年")){
                    salary=nf.format((Float.parseFloat(sa[0])+Float.parseFloat(sa[1]))/24)+"";
            }

            // 返回清洗后满足条件的一行数据
            context.write(new Text(city+","+title+","+salary+","+num+","+time),NullWritable.get());
        }
    };
  }
}
```

3. 打包执行

程序编写好之后，使用 Eclipse 的 export 来导出 jar 包，可以只选择要导出的类，操作步骤如下。

右击项目 → 选择 "export" → 选择 "java" → 选择 "jar file" → 单击 "next" → 选中 PreProcess 类 → 设置保存位置为 "/home/hadoop" → 设置 jar 的名称为 "PreProcess.jar"。

然后执行该 jar。

```
[hadoop@server1 ~]$ hadoop jar PreProcess.jar org.lanqiao.bigdata.PreProcess /input /out
```

4. 验证数据结果

执行完毕后，查看输出结果。

```
[hadoop@server1 ~]$ hadoop fs -cat /out/part*
北京,大数据开发工程师,1.67,5,2022-03-25
北京,大数据开发工程师,1.75,10,2022-03-23
北京,大数据架构师,1.8,1,2022-03-26
北京,大数据算法工程师,1.6,1,2022-03-24
北京,大数据运维工程师,2.33,2,2022-03-20
……
```

发现内容符合要求，数据清洗完毕。

2.6 任务五：使用 Hive 分析数据

使用 Hive 分析数据的流程如图 2.6 所示。

图 2.6 使用 Hive 分析数据的流程

按照操作步骤，可以将 Hive 分析数据的过程划分为 4 个阶段，分别是导入分析数据、分析需求、编写 HQL 语句和执行并验证结果。

清洗后的数据存储在 HDFS 上的文件 /out/part* 中，下面使用 Hive 来完成数据分析。

针对需求一：各城市大数据相关职位招聘人数（地区+人数）。

（1）导入分析数据。

开启 Hive 客户端，然后在 Hive 中创建数据库 t1，在 t1 数据库中创建内部表 a1，该表有 5 个字段，分别和数据中的 5 列保持对应，列之间用 ","分割，将数据文件 /out/part* 导入该表，操作如下。

```
[hadoop@server1 ~]$ hive
hive> create database t1;
hive> use t1;
hive> create table a1(citystring,titlestring,salaryfloat,numint,time string) row format delimited fields terminated by ",";
hive> load data inpath 'hdfs://server1:9000/out/part*' into table a1;
hive> select * from a1 limit 10;
……
```

（2）分析需求。

对于需求一，地区是第 1 列"city"，如"北京"，人数是第 4 列"num"，如 1。这里按照城市分组汇总人数即可。

（3）编写 HQL 语句。

思路：在 HDFS 上创建一个目录，把分析数据以覆盖的方式插入该目录。

在 HDFS 上创建目录：

```
[hadoop@server1 ~]$ hadoop fs -mkdir /user/hive/warehouse/a1
```

然后把分析数据以覆盖的方式插入该目录：

```
hive>insert overwrite directory 'hdfs://server1:9000/user/hive/warehouse/a1' select city as name,sum(num) as num from a1 group by city;
```

（4）执行并验证结果。

执行完毕之后，可以查看该目录下的文件内容是否正确。

```
[hadoop@server1 ~]$ hadoop fs -cat /user/hive/warehouse/a1/*
北京  4815
上海  4556
广州  4280
深圳  4690
武汉  2804
杭州  3025
成都  2353
西安  1704
```

针对需求二：大数据各相关职位招聘数量（职位 + 人数）。

（1）导入分析数据。

数据库和表之前已经建好，数据之前也已经导入完毕。

（2）分析需求。

对于需求二，职位是第 2 列 "title"，如 "大数据开发工程师"，人数是第 4 列 "num"，如 1。这里按照职位分组汇总人数即可。

（3）编写 HQL 语句。

思路：在 HDFS 上创建一个目录，把分析数据以覆盖的方式插入该目录。

在 HDFS 上创建目录：

```
[hadoop@server1 ~]$ hadoop fs -mkdir /user/hive/warehouse/a2
```

然后把分析数据以覆盖的方式插入该目录：

```
hive>insert overwrite directory 'hdfs://server1:9000/user/hive/warehouse/a2' select title as name,sum(num) as num from a1 group by title;
```

（4）执行并验证结果。

执行完毕之后，可以查看该目录下的文件内容是否正确。

```
[hadoop@server1 ~]$ hadoop fs -cat /user/hive/warehouse/a2/*
大数据开发工程师        19195
大数据分析工程师        3669
大数据运维工程师        423
大数据架构师            282
大数据算法工程师        1553
大数据项目经理          565
大数据其他相关职位      2540
```

针对需求三：大数据相关职位平均薪资（职位+薪资）。

（1）导入分析数据。

数据库和表之前已经建好，数据之前也已经导入完毕。

（2）分析需求。

对于需求三，职位是第 2 列 "title"，如 "大数据开发工程师"，薪资是第 3 列 "salary"，如 1.5，单位是万元/月。这里按照职位分组求薪资的平均值即可。

（3）编写 HQL 语句。

思路：在 HDFS 上创建一个目录，把分析数据以覆盖的方式插入该目录。

在 HDFS 上创建目录：

```
[hadoop@server1 ~]$ hadoop fs -mkdir /user/hive/warehouse/a3
```

然后把分析数据以覆盖的方式插入该目录：

```
hive>insert overwrite directory 'hdfs://server1:9000/user/hive/warehouse/a3' select title as name,round(avg(salary),2) as num from a1 group by title;
```

（4）执行并验证结果。

执行完毕之后，可以查看该目录下的文件内容是否正确。

```
[hadoop@server1 ~]$ hadoop fs -cat /user/hive/warehouse/a3/*
大数据开发工程师        2.12
大数据分析工程师        1.66
大数据架构师            2.81
大数据运维工程师        1.72
大数据其他相关职位      1.58
大数据算法工程师        2.21
大数据项目经理          2.52
```

2.7 任务六：使用 Sqoop 导出数据

使用 Sqoop 导出数据的流程如图 2.7 所示。

图 2.7　使用 Sqoop 导出数据的流程

按照操作步骤，可以将 Sqoop 导出数据的过程划分为 3 个阶段，分别是建库和建表、编写 Sqoop 命令导出数据及验证数据结果。

针对需求一：各城市大数据相关职位招聘人数（地区＋人数）。

（1）建库和建表。

首先登录 MySQL，数据库账号为 hive，密码为 hive，然后在 MySQL 中创建数据库 sqoop，然后在 sqoop 下创建数据表 a1。a1 含有 2 个字段：一个是 name，字符串类型；另一个是 num，整数类型。

```
[hadoop@server1 ~]$ mysql -u hive -p
Enter password:
```

输入密码"hive"。

```
musql> create database sqoop;
mysql> use sqoop;
mysql> create table a1(name varchar(100),num int);
```

（2）编写 Sqoop 命令导出数据。

使用 Sqoop 把 HDFS 上的数据导出到 MySQL 中。

```
[hadoop@server1 ~]$ sqoop export --connect jdbc:mysql://server1:3306/sqoop
--username hive --password hive --table a1 --export-dir /user/hive/warehouse/a1
--input-fields-terminated-by '\001' --num-mappers 1
```

（3）验证数据结果。

执行完毕后，查看 MySQL 中的数据是否正确。

针对需求二：大数据各相关职位招聘数量（职位+人数）。

（1）建库和建表。

首先登录 MySQL，然后在数据库 sqoop 下创建数据表 a2。a2 含有 2 个字段：一个是 name，字符串类型；另一个是 num，整数类型。

```
[hadoop@server1 ~]$ mysql -u hive -p
Enter password:
```

输入密码"hive"。

```
mysql> use sqoop;
mysql> create table a2(name varchar(100),num int);
```

（2）编写 Sqoop 命令导出数据。

使用 Sqoop 把 HDFS 上的数据导出到 MySQL 中。

```
[hadoop@server1 ~]$ sqoop export --connect jdbc:mysql://server1:3306/sqoop
--username hive --password hive --table a2 --export-dir /user/hive/warehouse/a2
--input-fields-terminated-by '\001' --num-mappers 1
```

（3）验证数据结果。

执行完毕后，查看 MySQL 中的数据是否正确。

针对需求三：大数据相关职位平均薪资（职位+薪资）。

（1）建库和建表。

首先登录 MySQL，然后在数据库 sqoop 下创建数据表 a3。a3 含有 2 个字段：一个是 name，字符串类型；另一个是 num，单精度浮点小数类型。

```
[hadoop@server1 ~]$ mysql -u hive -p
Enter password:
```

输入密码"hive"。

```
mysql> use sqoop;
mysql> create table a3(name varchar(100),num float);
```

（2）编写 Sqoop 命令导出数据。

使用 Sqoop 把 HDFS 上的数据导出到 MySQL 中。

```
[hadoop@server1 ~]$ sqoop export --connect jdbc:mysql://server1:3306/sqoop
--username hive --password hive --table a3 --export-dir /user/hive/warehouse/a3
--input-fields-terminated-by '\001' --num-mappers 1
```

（3）验证数据结果。

执行完毕后，查看 MySQL 中的数据是否正确。

2.8 任务七：Flask+ECharts 数据可视化

Flask+ECharts 数据可视化的流程如图 2.8 所示。

图 2.8　Flask+ECharts 数据可视化的流程

按照操作步骤，可以将 Flask+ECharts 数据可视化过程划分为 4 个阶段，分别是搭建 Flask 后台、编写接口代码、验证 JSON 数据和可视化页面。

针对需求一：使用柱状图展示各城市大数据相关职位招聘人数，按照人数降序排列。

（1）搭建 Flask 后台。

新建一个项目文件夹 Demo，然后在 Demo 下新建一个 a1.py 文件和一个 templates 文件夹，然后在 templates 文件夹中新建 a1.html 文件。

（2）编写接口代码。

接口文件 a1.py 代码如程序清单 2.3 所示。

程序清单 2.3
```
# 导入 pymysql 模块，用来连接数据库
import pymysql
# 从 flask 模块中导入 Flask 类和 render_template 函数
from flask import Flask,render_template
# 从 pyECharts 模块导入 options 选项，并缩写为 opts
from pyECharts import options as opts
# 从 pyECharts 模块导入柱状图类 Bar
from pyECharts.charts import Bar

# 打开数据库连接
db=pymysql.connect(host='server1',user='hive',password='hive',database='sqoop')
```

```python
# 使用 cursor() 方法创建一个游标对象 cursor
cursor=db.cursor()
# 使用 execute() 方法执行 SQL 查询
cursor.execute("SELECT name,num from a1 order by num desc")
# 使用 fetchall() 获取所有记录列表
data=cursor.fetchall()
# 从 data 中取出 name
name=[i[0] for i in data]
# 从 data 中取出 num 出现的次数
num=[i[1] for i in data]
# 关闭数据库连接
db.close()

# 创建一个 Flask 实例，并设置静态文件目录为 templates
app=Flask(__name__,static_folder="templates")

# 该函数用来配置柱状图的格式
def bar_base():
    # 列表用来存放数据
    data_pair=[]
    #zip 函数将 name、num、颜色一一对应，然后依次插入列表 data_pair
    for k,v,c in zip(name,num,['red','orange','yellow','green','blue','purple','black','#123456','#789123','gray']):
        data_pair.append(
            opts.BarItem(
                # 数据项名称
                name=k,
                # 单个数据项的数值
                value=v,
                # 图元样式配置项，我们只配置了颜色
                itemstyle_opts=opts.ItemStyleOpts(color=c)
            ))
    # 为柱状图添加 x 轴数据、x 轴名称、y 轴数据、y 轴名称，主标题、副标题
    c=(
        Bar()
        .add_xaxis(name)
        .add_yaxis("数量",data_pair,itemstyle_opts=opts.ItemStyleOpts(color='red'))
        .set_global_opts(title_opts=opts.TitleOpts(
            title="各城市大数据相关职位招聘人数"),
```

```
                    subtitle=" 招聘人数变化趋势 ",
                    xaxis_opts=opts.AxisOpts(name=" 城市信息 ",axislabel_opts=
{"interval":"0"}),
                    yaxis_opts=opts.AxisOpts(name=" 招聘人数 "),
            ))
    return c

@app.route("/")
def index():
    return render_template("a1.html")

@app.route("/a1")
def get_bar_chart():
    c=bar_base()
    #dump_options_with_quotes 函数可以将字典转换为 JSON
    return c.dump_options_with_quotes()

if __name__=="__main__":
app.run(host="0.0.0.0",port=8080)
```

（3）验证 JSON 数据。

启动 a1.py 程序。

```
python  a1.py
```

打开浏览器，输入地址"http://server1:8080/a1"，查看 JSON 数据部分内容是否如下。

```
"data":[
    {
            "name":" 北京 ",
            "value":4815,
            "itemStyle":{"color":"red"}
    },
    {
            "name":" 深圳 ",
            "value":4690,
            "itemStyle":{"color":"orange"}
    },
    {
            "name":" 上海 ",
            "value":4556,
            "itemStyle":{"color":"yellow"}
    },
    ……
]
```

（4）可视化页面。

前端文件 a1.html 代码如程序清单 2.4 所示。

程序清单 2.4

```html
<!DOCTYPEhtml>
<html>
<head>
    <metacharset="UTF-8">
    <title>pyecharts</title>
    <!-- 引入 jquery-->
    <scriptsrc="https://lf26-cdn-tos.bytecdntp.com/cdn/expire-1-M/jquery/3.6.0/jquery.min.js"></script>
    <!-- 引入 echarts.js-->
    <scripttype="text/javascript"src="echarts.min.js"></script>
</head>
<body>
    <!-- 为 ECharts 准备一个特定大小（宽高）的图表显示区域 -->
    <divid="bar"style="width:1400px; height:640px;"></div>
    <script>
        $(
            function () {
                var chart=echarts.init(document.getElementById('bar'),'white',{ renderer:'canvas' });
                $.ajax({
                    type:"GET",
                    url:"/a1",
                    dataType:'json',
                    success:function(result){
                        chart.setOption(result);
                    }
                });
            }
        )
    </script>
</body>
</html>
```

代码编写好后，重新启动 a1.py 程序，打开浏览器，输入地址"http://localhost:8080"，就能看到可视化数据了，如图 2.9 所示。

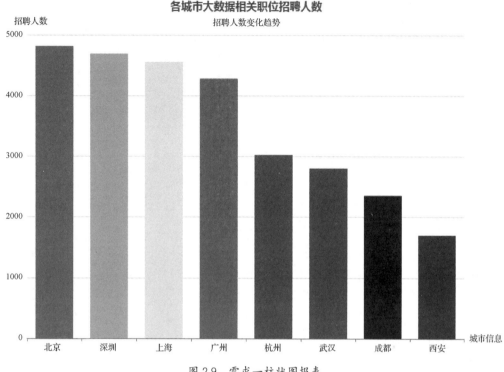

图 2.9 需求一柱状图报表

针对需求二：使用折线图展示大数据各相关职位招聘数量，按照人数降序排列。

（1）搭建 Flask 后台。

在需求一的基础上，在项目文件夹 Demo 下新建一个 a2.py 文件，然后在 templates 文件夹中新建 a2.html 文件。

（2）编写接口代码。

接口文件 a2.py 代码如程序清单 2.5 所示。

程序清单 2.5
```
# 导入 pymysql 模块，用来连接数据库
import pymysql
# 从 flask 模块中导入 Flask 类和 render_template 函数
from flask import Flask,render_template
# 从 pyecharts 模块导入 options 选项，并缩写为 opts
from pyecharts import options as opts
# 从 pyecharts 模块导入折线图类 Line
from pyecharts.charts import Line
```

```python
# 打开数据库连接
db=pymysql.connect(host='server1',user='hive',password='hive',database='sqoop')
# 使用 cursor() 方法创建一个游标对象 cursor
cursor=db.cursor()
# 使用 execute() 方法执行 SQL 查询
cursor.execute("SELECT name,num from a2 order by num desc")
# 使用 fetchall() 获取所有记录列表
data=cursor.fetchall()
# 从 data 中取出 name
name=[i[0] for i in data]
# 从 data 中取出 num 出现的次数
num=[i[1] for i in data]
# 关闭数据库连接
db.close()

# 创建一个 Flask 实例，并设置静态文件目录为 templates
app=Flask(__name__,static_folder="templates")

# 该函数用来配置折线图的格式
def line_base():
    # 列表用来存放数据
    data_pair=[]
    #zip 函数将 name、num、颜色一一对应，然后依次插入列表 data_pair
    for k,v,c in zip(name,num,['red','orange','yellow','green','blue','purple','black','#123456','#789123','gray']):
        data_pair.append(
            opts.LineItem(
                # 数据项名称
                name=k,
                # 单个数据项的数值
                value=v,
                # 图元样式配置项，我们只配置了颜色
                itemstyle_opts=opts.ItemStyleOpts(color=c)
            ))
    # 为折线图添加 x 轴数据、x 轴名称、y 轴数据、y 轴名称，主标题、副标题
    c=(
        Line()
        .add_xaxis(name)
        .add_yaxis("数量",data_pair,itemstyle_opts=opts.ItemStyleOpts(color='red'))
```

```python
            .set_global_opts(title_opts=opts.TitleOpts(
                title=" 大数据各相关职位招聘数量 "),
                subtitle=" 招聘数量变化趋势 ",
                xaxis_opts=opts.AxisOpts(name=" 职位名称 ",axislabel_opts={"interval":"0"}),
                yaxis_opts=opts.AxisOpts(name=" 职位数量 "),
                ))
    return c

@app.route("/")
def index():
    return render_template("a2.html")

@app.route("/a2")
def get_line_chart():
    c=line_base()
    #dump_options_with_quotes 函数可以将字典转换为 JSON
    return c.dump_options_with_quotes()

if __name__ == "__main__":
    app.run(host="0.0.0.0",port=8080)
```

(3) 验证 JSON 数据。

启动 a2.py 程序。

```
python a2.py
```

打开浏览器,输入地址"http://server1:8080/a2",查看 JSON 数据部分内容是否如下。

```
"data":[
    {
        "name":" 大数据开发工程师 ",
        "value":19195,
        "itemStyle":{"color":"red"}
    },
    {
        "name":" 大数据分析工程师 ",
        "value":3669,
        "itemStyle":{"color":"orange"}
    },
    {
        "name":" 大数据其他相关职位 ",
        "value":2540,
```

```
            "itemStyle":{"color":"orange"}
        },
        ......
]
```

（4）可视化页面。

前端文件 a2.html 代码如程序清单 2.6 所示。

程序清单 2.6

```html
<!DOCTYPE html>
<html>
<head>
    <metacharset="UTF-8">
    <title>pyecharts</title>
    <!-- 引入 jquery-->
    <scriptsrc="jquery.min.js"></script>
    <!-- 引入 echarts.js-->
    <scripttype="text/javascript"src="echarts.min.js"></script>
</head>
<body>
    <!-- 为 ECharts 准备一个特定大小（宽高）的图表显示区域 -->
    <divid="line"style="width:1400px; height:640px;"></div>
    <script>
        $(
            function () {
                var chart=echarts.init(document.getElementById('line'),'white',{ renderer:'canvas' });
                $.ajax({
                    type:"GET",
                    url:"/a2",
                    dataType:'json',
                    success:function(result){
                        chart.setOption(result);
                    }
                });
            }
        )
    </script>
</body>
</html>
```

代码编写好后，重新启动 a2.py 程序，打开浏览器，输入地址"http://localhost:8080"，就能看到可视化数据了，如图 2.10 所示。

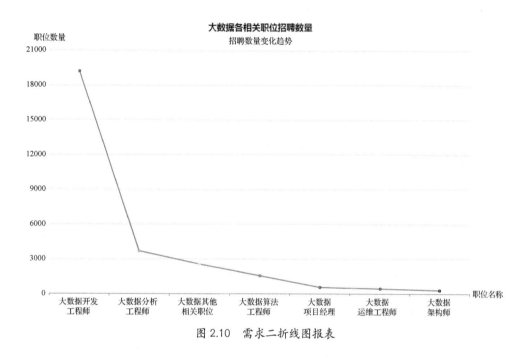

图 2.10　需求二折线图报表

针对需求三：使用柱状图展示大数据相关职位平均薪资，按照平均薪资降序排列。

（1）搭建 Flask 后台。

在需求二的基础上，在项目文件夹 Demo 下新建一个 a3.py 文件，然后在 templates 文件夹中新建 a3.html 文件。

（2）编写接口代码。

接口文件 a3.py 代码如程序清单 2.7 所示。

程序清单 2.7
```
# 导入 pymysql 模块，用来连接数据库
import pymysql
# 从 flask 模块中导入 Flask 类和 render_template 函数
from flask import Flask,render_template
# 从 pyecharts 模块导入 options 选项，并缩写为 opts
from pyecharts import options as opts
# 从 pyecharts 模块导入柱状图类 Bar
from pyecharts.charts import Bar
```

```python
# 打开数据库连接
db=pymysql.connect(host='server1',user='hive',password='hive',database='sqoop')
# 使用 cursor() 方法创建一个游标对象 cursor
cursor=db.cursor()
# 使用 execute() 方法执行 SQL 查询
cursor.execute("SELECT name,num from a3 order by num desc")
# 使用 fetchall() 获取所有记录列表
data=cursor.fetchall()
# 从 data 中取出 name
name=[i[0] for i in data]
# 从 data 中取出 num 出现的次数
num=[i[1] for i in data]
# 关闭数据库连接
db.close()

# 创建一个 Flask 实例,并设置静态文件目录为 templates
app=Flask(__name__,static_folder="templates")

# 该函数用来配置柱状图的格式
def bar_base():
    # 列表用来存放数据
    data_pair=[]
    #zip 函数将 name、num、颜色一一对应,然后依次插入列表 data_pair
    for k,v,c in zip(name,num,['red','orange','yellow','green','blue','purple','black','#123456','#789123','gray']):
        data_pair.append(
            opts.BarItem(
                # 数据项名称
                name=k,
                # 单个数据项的数值
                value=v,
                # 图元样式配置项,我们只配置了颜色
                itemstyle_opts=opts.ItemStyleOpts(color=c)
            ))
    # 为柱状图添加 x 轴数据、x 轴名称、y 轴数据、y 轴名称、主标题、副标题
    c=(
        Bar()
        .add_xaxis(name)
        .add_yaxis(" 数量 ",data_pair,itemstyle_opts=opts.ItemStyleOpts(color='red'))
```

```python
            .set_global_opts(title_opts=opts.TitleOpts(
                title="大数据相关职位平均薪资"),
                subtitle="职位薪资差别",
                xaxis_opts=opts.AxisOpts(name="职位名称",axislabel_opts=
{"interval":"0"}),
        yaxis_opts=opts.AxisOpts(name="平均薪资(万元/月)"),
                ))
    return c

@app.route("/")
def index():
    return render_template("a3.html")

@app.route("/a3")
def get_bar_chart():
    c=bar_base()
    #dump_options_with_quotes 函数可以将字典转换为 JSON
    return c.dump_options_with_quotes()

if __name__=="__main__":
    app.run(host="0.0.0.0",port=8080)
```

(3) 验证 JSON 数据。

启动 a3.py 程序。

```
python  a3.py
```

打开浏览器,输入地址"http://server1:8080/a3",查看 JSON 数据部分内容是否如下。

```
"data":[
    {
            "name":"大数据架构师",
            "value":2.81,
            "itemStyle":{"color":"red"}
    },
    {
            "name":"大数据项目经理",
            "value":2.52,
            "itemStyle":{"color":"orange"}
    },
    {
            "name":"大数据算法工程师",
            "value":2.21,
```

```
            "itemStyle":{"color":"yellow"}
        },
        ......
]
```

（4）可视化页面。

前端文件 a3.html 代码如程序清单 2.8 所示。

程序清单 2.8
```
<!DOCTYPE html>
<html>
<head>
    <metacharset="UTF-8">
    <title>pyecharts</title>
    <!-- 引入 jquery-->
    <scriptsrc="jquery.min.js"></script>
    <!-- 引入 echarts.js-->
    <scripttype="text/javascript"src="echarts.min.js"></script>
</head>
<body>
    <!-- 为ECharts准备一个特定大小（宽高）的图表显示区域 -->
    <divid="bar"style="width:1400px; height:640px;"></div>
    <script>
        $(
            function () {
                var chart=echarts.init(document.getElementById('bar'),'white',{ renderer:'canvas' });
                $.ajax({
                    type:"GET",
                    url:"/a3",
                    dataType:'json',
                    success:function(result){
                        chart.setOption(result);
                    }
                });
            }
        )
    </script>
</body>
</html>
```

代码编写好后，重新启动 a3.py 程序，打开浏览器，输入地址"http://localhost:8080"，就能看到可视化数据了，如图 2.11 所示。

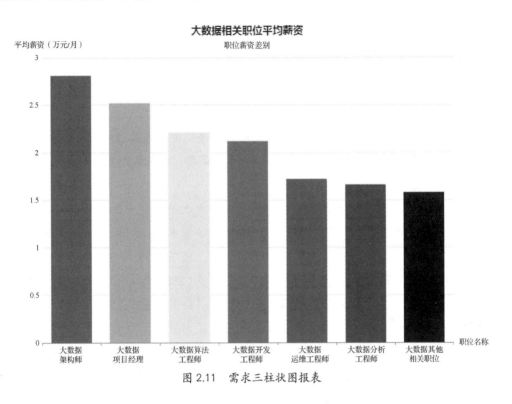

图 2.11 需求三柱状图报表

2.9 任务八：编写分析报告

1. 请结合数据分析结果或现实观察回答以下问题

问题1：根据分析结果，大数据相关职位需要哪些主要技能？为什么？

问题2：根据分析结果，各地大数据产业发展情况如何？

问题3：根据市场需求分析，大数据行业有哪些人才培养方向？为什么？

问题4：请简述今后大数据发展方向在哪里。

2. 问题参考答案

问题1：根据需求二的分析结果，可以得出大数据岗位所需要的技能包括开发、数据分析、算法、运维、项目经验、架构等。其中排名第一的是大数据开发（大数据开发工程师：19195人），因为首先搭建大数据平台，然后在平台上进行二次开发，比如在此基础上构建数据分析平台、算法平台，以及后期的维护升级，都需要开发的支持，所以对大数据开发工程师的需求量最大。数据分析（大数据分析工程师：3669人）是排名第二的需求，大数据要发挥价值，主要依靠大数据的分析结果，需要通过统计分析，得出有价值的结论，指导决策。排名第三的是算法（大数据算法工程师：1553人），有些问题仅仅使用数据分析是搞不定的，需要使用算法来支持，比如预测下一个数据，这部分需求对能力的要求更高，所以在招的人数要少一些。项目经验（大数据项目经理：565人）排名第四，行业对具备较强开发能力和丰富业务经验的人的需求较少。运维（大数据运维工程师：423人）排名第五，大数据项目的日常运行需要运维人员的参与，这部分需求少，但是必不可少。架构（大数据架构师：282人）排名最后，有丰富经验的架构师必不可少，但是需求较少。

问题2：根据需求一的分析结果，我们可以发现大数据产业发展和地域的关系非常紧密，一线城市、沿海城市、东部城市发展得比较好，公司多，岗位多，工资高；一般来说，产业发展情况和经济挂钩，比如北京（4815人）、上海（4556人）、深圳（4690人）、广州（4280人）产业发展最好；中部区域其次，如武汉（2804人）、西安（1704人）等地；西部区域最差，主要集中在成都（2353人）、重庆、贵州等城市。

问题3：根据需求二的分析结果，大数据行业的人才培养方向主要有大数据开发工程师、大数据分析师、大数据算法工程师、大数据挖掘工程师、大数据运维工程师、大数据产品工程师、大数据售前工程师、大数据系统架构师等。人才培养方向主要和需求挂钩，大数据开发工程师和分析师是需求最大的。

问题4：在全国一体化大数据中心体系建设部署下，以数据中心为代表的数字基础设施将加快突破，在区域协同、数据流通、算力普惠等方面形成合力，进一步增强我国大数据产业发展基础优势。数据中心统筹推进、一体建设态势明显，（超）大型数据中心规模化建设布局持续优化，京津冀、长三角、粤港澳大湾区等8个全国一体化算力网络国家枢纽节点领先发展，在区域数据中心集群建设、算力协同调度机制突破、市场化运营模式创新等方面率先突破。"东数西算"工程全面实施，在突出地区优势、补足发展短板的原则下，将推动东西部地区数据中心差异化、协同化建设。

2.10 答疑解惑

问题1：项目中使用Python"爬取"招聘网站数据时可能遇到哪些问题？该如何解决？

解答1：

（1）JS加密技术。招聘网站对部分JS代码采取了加密的手段，来保护自己的代码。对于"爬虫"用户来说，"爬取"的乱码不能直接使用，需要解密处理。可以通过debug找到JS加密、解密的代码，然后通过Python重新实现，但是这种方法很耗时；或者使用Selenium工具来实现，不过效率较低。

（2）"爬虫"被封禁。"爬虫"经常被禁用，一般来说，需要注意以下事项：遵守法律法规，构造合理的http访问头（如user-agent的值），放慢"爬虫"的访问速度，采用动态IP地址，保证"爬虫"访问路径和用户访问路径一致等。这里强调一下法律法规问题，从技术中立的角度来看，"爬虫"技术本身并无违法违规之处，使用"爬虫"技术是否触犯法律底线，取决于主体如何使用、为何使用。比如是否"爬取"了个人信息，侵犯了公民个人隐私；是否"爬取"了商业秘密，造成不正当竞争行为；是否"爬取"了作品信息，侵犯了著作权。同时要满足网站的robots协议，避免对网站的使用造成影响。

（3）"爬虫"中的去重。在"爬取"网站内容时，有的内容很可能经过了多次引用或者转载，导致多个地址上有着相同的内容。如果不处理的话，会获得大量重复的内容。解决办法是选取具有区分度的标识符来判断特定内容是否已经被"爬取"过，从而达到去重的效果。

问题2： 项目中使用Hive时遇到如下错误，该如何解决？

```
Caused by:org.datanucleus.store.rdbms.connectionpool.DatastoreDriverNotFound
Exception:The specified datastore driver ("com.mysql.jdbc.Driver") was not found
in the CLASSPATH. Please check your CLASSPATH specification,and the name of the
driver.
   atorg.datanucleus.store.rdbms.connectionpool.AbstractConnectionPoolFactory.
loadDriver(AbstractConnectionPoolFactory.java:58)
   at org.datanucleus.store.rdbms.connectionpool.BoneCPConnectionPoolFactory.
createConnectionPool(BoneCPConnectionPoolFactory.java:54)
   at org.datanucleus.store.rdbms.ConnectionFactoryImpl.generateDataSources
(ConnectionFactoryImpl.java:213)
```

解答2： 上述问题是缺少MySQL的驱动包所导致的，解决方法是下载mysql-connector-java-5.1.35.tar.gz，然后复制到Hive的lib目录下。

```
shiyanlou:~/ $ cp mysql-connector-java-5.1.35-bin.jar /opt/hive-1.1.0/lib/
```

问题 3：项目中启动 Hive 时，无法进入客户端，报如下错误，该如何解决？

```
    Exception in thread "main" java.lang.RuntimeException:Hive metastore database 
is not initialized. Please use schematool (e.g. ./schematool -initSchema 
-dbType ...) to create the schema. If needed,don't forget to include the option 
to auto-create the underlying database in your JDBC connection string (e.g. 
?createDatabaseIfNotExist=true for mysql)
```

解答 3：出现该问题是由于没有初始化元数据。运行 schematool –dbTypemysql –initSchema 命令进行 Hive 元数据库的初始化。

```
shiyanlou:~/ $ schematool -dbTypemysql -initSchema
```

问题 4：项目中 Hive 应该使用内部表还是外部表？

解答 4：内部表与外部表的区别如下。

（1）内部表数据由 Hive 自身管理，外部表数据由 HDFS 管理。

（2）内部表数据存储的位置默认是 /user/hive/warehouse，外部表数据的存储位置由自己指定。

（3）删除内部表会直接删除元数据以及存储数据，删除外部表仅仅会删除元数据，HDFS 上的文件不会被删除。

实践中通常使用外部表，简单易操作，数据无须移动。

问题 5：执行 Hive 时经常会遇到数据分布不均的情况，有哪些原因可能会造成数据分布不均？

解答 5：主要有以下几个原因。

（1）key 分布不均匀。

（2）业务数据本身的特性。

（3）建表考虑不周全。

（4）某些 HQL 语句本身就存在数据分布不均的情况。

问题 6：如何使用 Hive 来完成 WordCount 功能？

解答 6：

（1）准备文本内容。

新建一个文本文件 wc.txt，内容如下所示。

```
shiyanlou:~/ $ cat wc.txt
aaa  bbb  bbb
aaa  ccc  ddd  bbb
```

（2）新建 Hive 表。

```
hive> create table wc(line string);
```

没有指定列的分隔符，因为只有一列，行数据整体是二维结构中的单元格内容。

（3）加载数据到 Hive 表中。

```
hive> load data local inpath '/home/shiyanlou/wc.txt' into table wc;
loading data to table t1.wc
hive> select * from wc;
aaa  bbb  bbb
aaa  ccc  ddd  bbb
```

（4）分割文本。

```
hive> select split(line,'\\s') as word from wc;
["aaa","bbb","bbb"]
["aaa","ccc","ddd","bbb"]
```

这里使用了 Hive 的内置函数 split 来切割每一行的内容，使用正则表达式 \s，即空白符（包括空格、tab 缩进、换行等所有的空白形式）来分割，多出的一个"\"是用来转义的。可以看出分割后呈现为字符串数组的形式。

（5）行转列。

上一步分割完成后单词仍然在同一行，无法统计，需要行转列。

```
hive> select explode(split(line,'\\s')) as word from wc;
aaa
bbb
bbb
aaa
ccc
ddd
bbb
```

使用 Hive 的内置函数 explode 可以做到行转列，现在每个单词都单独占据一行。

（6）统计计数。

```
hive> select word,count(*) as count from (select explode(split(line,'\\s')) as word from wc) w group by word order by count desc;
……
bbb 3
aaa 2
ccc 1
ddd 1
```

至此，Hive 已实现 WordCount 计数功能。

2.11 拓展练习

1．拓展项目

我们拓展一下本章项目，假设有以下需求。

按要求使用折线图展示各城市大数据相关职位平均薪资，并在前端显示。具体要求如下所示。

主标题：各城市大数据相关职位平均薪资。

副标题：地区薪资差别。

横坐标：城市信息。

纵坐标：平均薪资（万元／月）。

按照本章所讲的步骤来做。请同学们自行完成该任务。

2．强化 Hive 语句

数据格式如下：

```
2010012325
```

2010012325 表示 2010 年 1 月 23 日的最高气温为 25℃。

现有一批初始数据：

```
2014010114
2014010216
2014010317
2014010410
2014010506
2012010609
2012010732
2012010812
2012010919
2012011023
2001010116
2001010212
2001010310
2001010411
2001010529
2013010619
2013010722
2013010812
```

```
2013010929
2013011023
2008010105
2008010216
2008010337
2008010414
2008010516
2007010619
2007010712
2007010812
2007010999
2007011023
2010010114
2010010216
2010010317
2010010410
2010010506
2015010649
2015010722
2015010812
2015010999
2015011023
```

请编写 HQL 语句，求每年的最高温度，格式如：2015   99。

第 3 章
电商网站实时数据分析项目

项目标签：Spark 实时分析

【本章简介】

本章以"电商网站实时数据分析项目"的推进为主线来介绍大数据实时计算项目的详细过程，力求帮助读者掌握大数据实时计算的过程及所需工具、技术。本章首先分析项目的需求和技术架构，帮助读者了解项目目标，厘清大体思路。接着按照大数据实时计算的流程分步骤详细介绍数据采集、数据缓存、数据计算、数据存储、数据可视化等操作。最后通过答疑解惑和拓展练习帮助读者巩固知识，强化技能。

【知识地图】

本章项目涉及的知识地图如图 3.1 所示，读者在学习过程中可以查询相关知识卡片，补充理论知识。

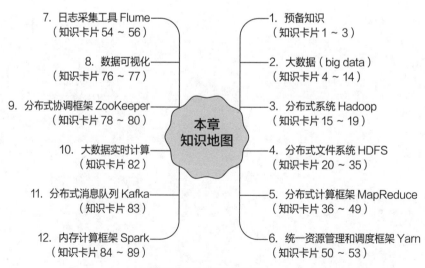

图 3.1 本章知识地图

3.1 任务一：需求分析

1．业务背景

电商网站的数据非常多，包括用户基本数据（个人信息、账号信息等），用户的访问数据（浏览、点击、收藏、购买等），以及用户的消费数据（支付、转账等）。借助不同的数据，可以从不同的角度分析用户，这里我们以用户的消费数据为例，重点来看如何分析。

用户通过支付系统购买商品后，就会立刻生成一笔订单，该订单含有非常多的信息，包括商品名称、商品单价、商品数量、订单金额、商品类别、卖家名称、购买地址、购买时间、订单号、用户编号、卖家编号等信息，而且一笔订单里包含的商品也可能有多种。

不同的用户有着各自的需求，比如领导可能只关心销售总额和订单量，部门经理可能只关心他所负责的某条业务线的销售情况，而卖家则只关心店铺的销售情况。可见，不同的人所关心的指标不太一样，关心的业务范围也不一样，但是相同点是都需要实时了解数据情况。

由于订单数据指标非常多，这里我们简化一下，保留少许用得上的指标，订单的格式如下所示。

```
{'id':'101','info':[{'name':'可乐','price':2.00,'num':3,'sum':6.00,'type':'饮料'},
          {'name':'辣条','price':1.00,'num':5,'sum':5.00,'type':'食品'}],
'time':'2022-04-18 12:13:14'}
  {'id':'102','info':[{'name':'洗衣机','price':2500.00,'num':1,'sum':2500.00,'type':
'家电'}],
       'time':'2022-04-18 14:20:01'}
  ……
```

其中每笔订单就是一个 JSON 字符串，含有 3 个属性。"id"属性表示订单号。"info"属性表示具体商品的信息，是一个数组，里面的每一个对象描述的都是一种商品。比如"101"这个订单含有 2 种商品，一种商品是可乐，单价为 2 元，数量为 3，金额为 6 元，类型为饮料；另一种商品是辣条，单价为 1 元，数量为 5，金额为 5 元，类型为食品。"102"这个订单含有 1 种商品，即洗衣机，单价为 2500 元，数量为 1，金额为 2500 元，类型为家电。"time"属性表示订单成交时间。

2．业务需求

（1）企业高层希望能够随时查看当天的累计销售总额，这样可以和之前的情况做对比，同时希望知晓异常数据，比如大幅增加或者减少的原因。

（2）部门经理希望能够随时查看某种类型商品（如家电、食品）当天每个时间段（如每个小时）内的销售金额，以便观察日内销售情况的变化。

3．功能需求

（1）统计当天的累计销售金额，使用 ECharts 中的折线图，实时动态地显示。

（2）统计某种类型的商品当天每个时间段内的销售金额，使用 ECharts 中的折线图，实时动态地显示。

3.2　任务二：项目方案设计

1．技术领域判定

根据需求分析，我们知道该项目有以下几个特征。

（1）数据量大。

（2）需要实时分析。

（3）需要形成简单的报表。

这是一个对用户消费数据进行实时统计分析的场景，解决方案是典型的大数据实时计算技术。

2．技术选型

大数据实时计算的行业最佳实践架构由下往上分为 5 层：数据采集层、数据缓存层、数据计算层、数据存储层、数据展示层。下面我们对这 5 层进行技术选型。

（1）数据采集层。我们需要的订单数据存储在服务器上，服务器通常会提供数据访问下载的接口，用户可以通过远程的方式来采集。可以使用的技术方案有 Flume、Scribe 等。其中 Flume 是一种纯粹为数据采集而生的分布式服务，我们这里选用 Flume。

（2）数据缓存层。采集来的数据通常先存储在数据中间体中，方便消息传递和消息消费。可以使用的技术方案有 ActiveMQ、RabbitMQ、Kafka 等。其中 Kafka 可以实时处理大量数据，满足各种需求场景，这里我们选用 Kafka。

（3）数据计算层。该层技术较多，可以使用的技术有 Spark、Storm、Flink 等。考虑到数据计算的简单性，以及技术的普及性，我们选用已经非常流行和稳定的 Spark 作为计算框架。

（4）数据存储层。在实时计算领域，通常使用 Redis 这种基于内存的数据库来提高速度，这种数据库可以分布式部署、使用，适合大数据领域的实时存储。

（5）数据展示层。展示层技术较多，可根据技术团队的技术基础自行选择，这里我们选择 Java 技术栈里较为主流的 SpringBoot+SSM（由 Spring+SpringMVC+MyBatis 框架整合而成，常作为 Java Web 项目的框架）作为后端技术，结合 ECharts 等前端技术，以图形化的方式来展示数据结果，如饼图、柱状图、折线图等。

3．实现流程

本项目的流程如图 3.2 所示，首先使用 Flume 采集服务器上的用户消费订单数据，存储到 Kafka 消息中间件中，然后使用 Spark 从 Kafka 中提取数据进行实时计算，并把计算结果存储到 Redis 数据库中，再创建 SpringBoot 应用程序来访问 Redis 数据库，最后使用 ECharts 将数据可视化呈现在网页上。

图 3.2　大数据实时计算流程

4．开发环境准备

在完成后续任务之前，我们需要安装并配置一系列软件，准备好开发环境，如表 3.1 所示。这些基础操作，我们放入附录中以供不熟悉的读者参考。

表 3.1　开发环境准备

操作	附录
Hadoop 安装部署和配置	附录 1
Flume 安装部署和配置	附录 5
Kafka 安装部署和配置	附录 10
Spark 安装部署和配置	附录 11

3.3　任务三：使用 Flume+Kafka 实时收集数据

使用 Flume+Kafka 实时收集数据的流程如图 3.3 所示。

图 3.3　使用 Flume+Kafka 实时收集数据的流程

按照操作步骤,可以将 Flume+Kafka 实时收集数据的过程划分为 4 个阶段,分别是创建采集数据文件、创建采集配置文件、启动 Flume 和 Kafka 以及验证采集结果。

1．创建采集数据文件

在 ~ 目录下新建文件 data.log,该文件用来模拟产生交易订单日志数据(我们会向文件中写入交易数据)。

```
[hadoop@master~]$ touch data.log
```

2．创建采集配置文件

在 ~ 目录下新建 Flume 的配置信息文件 exec.conf,完整代码如程序清单 3.1 所示。

程序清单 3.1

```
[hadoop@master~]$ vi exec.conf
a1.sources=r1
a1.channels=c1
a1.sinks=k1

a1.sources.r1.type=exec
a1.sources.r1.command=tail -F /home/hadoop/data.log
a1.sources.r1.channels=c1

a1.channels.c1.type=memory
a1.channels.c1.capacity=10000
a1.channels.c1.transactionCapacity=100

a1.sinks.k1.type=org.apache.flume.sink.kafka.KafkaSink
# 设置 Kafka 的主题 topic
a1.sinks.k1.topic=order
# 设置消费者编码为 UTF-8
a1.sinks.k1.custom.encoding=UTF-8
# 绑定 Kafka 主机及端口号
a1.sinks.k1.brokerList=192.168.128.131:9092,192.168.128.132:9092,192.168.128.133:9092
# 设置 Kafka 序列化方式
```

```
a1.sinks.k1.serializer.class=kafka.serializer.StringEncoder
a1.sinks.k1.requiredAcks=1
a1.sinks.k1.batchSize=20
a1.sinks.k1.channel=c1
```

3．启动 Flume 和 Kafka 服务

启动 Flume 服务：

```
[hadoop@master soft]$ flume-ng agent -n a1 -f /home/hadoop/exec.conf -Dflume.root.logger=INFO,console
```

启动 Kafka 服务：

```
[hadoop@master soft]$ kafka-server-start.sh -daemon $KAFKA_HOME/config/server.properties
```

4．验证采集结果

在 Kafka 中创建一个名为 order 的主题。

```
[hadoop@master soft]$ kafka-topics.sh --create --zookeeper 192.168.128.131:2181 --replication-factor 2 --partitions 3 --topic order
```

启动 Kafka 消费者客户端，准备消费主题 order 的消息。

```
[hadoop@slave1 soft]$ kafka-console-consumer.sh --bootstrap-server 192.168.128.131:9092 --from-beginning --topic order --group ppp
```

使用 >> 向 /home/hadoop/data.log 文件中追加内容。

```
[hadoop@master ~]$ echo 'hello' >> data.log
```

可以看到该条消息被 Kafka 消费者消费了。

至此，Flume 和 Kafka 的配置已经完成。

3.4 任务四：使用 Spark 实时计算数据

本任务涉及 Spark RDD 的操作，对此不熟悉的读者可以参考表 3.2 中提到的相关附录进行学习。

表 3.2　Spark RDD 操作

操作	附录
Spark RDD 算子	附录 12
通过 WordCount 熟悉 Spark RDD	附录 13

使用 Spark 实时计算数据的流程如图 3.4 所示。

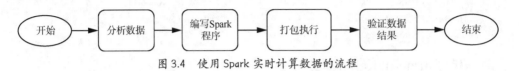

图 3.4 使用 Spark 实时计算数据的流程

按照操作步骤，可以将 Spark 实时计算数据的过程划分为 4 个阶段，分别是分析数据、编写 Spark 程序、打包执行和验证数据结果。

针对需求一：统计当天的累计销售金额，使用 ECharts 中的折线图，实时动态地显示。

（1）分析数据。

累计销售金额实际上是每一种商品金额的和。一笔订单可能有多件商品，需要把所有商品的金额累加起来。

（2）编写 Spark 程序。

由于订单数据是 JSON 格式，所以需要使用 JSON 解析包，这里使用阿里巴巴的 fastjson。计算结束后，还需要把结果存储到 Redis 中。

首先引入相关依赖。

```xml
<dependency>
    <groupId>com.alibaba</groupId>
    <artifactId>fastjson</artifactId>
    <version>1.2.73</version>
</dependency>
<dependency>
    <groupId>redis.clients</groupId>
    <artifactId>jedis</artifactId>
    <version>3.2.0</version>
</dependency>
```

然后编写 Spark 代码计算累计销售金额，完整代码如程序清单 3.2 所示。

程序清单 3.2

```
package org.lanqiao.bigdata

import java.util.HashSet

import org.apache.kafka.clients.consumer.ConsumerConfig
import org.apache.kafka.common.serialization.StringDeserializer
import org.apache.spark.SparkConf
```

```scala
import org.apache.spark.streaming.Seconds
import org.apache.spark.streaming.StreamingContext
import org.apache.spark.streaming.kafka010.ConsumerStrategies
import org.apache.spark.streaming.kafka010.KafkaUtils
import org.apache.spark.streaming.kafka010.LocationStrategies

import com.alibaba.fastjson.JSON

import redis.clients.jedis.HostAndPort
import redis.clients.jedis.JedisCluster

object SparkOrder {
  def main(args:Array[String]):Unit={
    valsparkconf=new SparkConf().setAppName("sparkorder")
    //设置 DStream 批次时间间隔为 5 秒
    valssc=new StreamingContext(sparkconf,Seconds(5))
    //设置 Kafka 的配置：zk 集群链接、消费组
    valkafkaParams=Map[String,Object](
      ConsumerConfig.BOOTSTRAP_SERVERS_CONFIG->"192.168.128.131:9092,192.168.128.132:9092,192.168.128.133:9092",
      ConsumerConfig.GROUP_ID_CONFIG ->"mygroup",
      ConsumerConfig.KEY_DESERIALIZER_CLASS_CONFIG
        ->classOf[StringDeserializer],
      ConsumerConfig.VALUE_DESERIALIZER_CLASS_CONFIG
        ->classOf[StringDeserializer])
    //设置消费主题
    val topics=Array("order")
    //获取 Kafka 数据流
    val stream=KafkaUtils.createDirectStream[String,String](
      ssc,
      LocationStrategies.PreferConsistent,
      ConsumerStrategies.Subscribe[String,String](topics,kafkaParams))
    //获取每一行的数据
    val line=stream.map(_.value)
    //val line=stream.map(_._2)
    line.print()
    // {'id':'101','info':[{'name':'可乐','price':2.00,'num':3,'sum':6.00,'type':'饮料'},
    //                     {'name':'辣条','price':1.00,'num':5,'sum':5.00,'type':'食品'}],
    // 'time':'2022-04-18 12:13:14'}
    //连接 Redis 集群
```

```
valjedisClusterNodes=new HashSet[HostAndPort]
jedisClusterNodes.add(new HostAndPort("192.168.128.131",6379))
jedisClusterNodes.add(new HostAndPort("192.168.128.132",6379))
jedisClusterNodes.add(new HostAndPort("192.168.128.133",6379))
val jedis=new JedisCluster(jedisClusterNodes)
    // 解析数据
val data=line.map { a =>{
    val json=JSON.parseObject(a)
    val info=json.getJSONArray("info")
    var sum=0.0
    for (i<-0 until info.size()){
        sum+=info.getJSONObject(i).getDouble("sum")
    }
    // 累计求和并存入 Redis
    var orderSum=jedis.incrByFloat("orderSum",sum)
    // 求和后,打印累计的返回值
    print(" 累计销售金额: "+orderSum)
}}

    // 开始计算
ssc.start()
    // 等待停止
ssc.awaitTermination()
    // 关闭 Jedis
jedis.close()
    }
}
```

程序清单 3.2 的大体思路如下。

① 连接 Kafka 服务器,每隔 5 秒尝试消费 order 主题下的订单消息,这里获取的是自上一次消费以来的 5 秒内的若干条订单数据。

② 对于每条订单数据,解析出订单金额 orderSum,并将其累加至 Redis 中的 orderSum 变量中。

③ 最后关闭相关资源。

(3) 打包执行。

将程序打成 jar 包 SparkOrder.jar,上传到 Spark 集群,提交 Spark 任务。

```
[hadoop@master ~]$ spark-submit --master yarn --deploy-mode cluster \
--class org.lanqiao.bigdata.SparkOrder --executor-memory 1g \
/home/hadoop/SparkOrder.jar
```

启动 Spark Streaming 后,我们编写的程序就能获取 order 主题下的消息,解析 JSON 格式的消息并执行计算,将计算结果存储到 Redis 数据库中。

(4)验证数据结果。

我们可以到 Yarn 的 8088 端口上找到对应任务,然后去查看任务的 log 日志,程序中的输出将会体现在日志中。

当我们使用 >> 往 /home/hadoop/data.log 文件中追加如下订单时:

```
[hadoop@master ~]$ echo "{'id':'101','info':[{'name':'可乐','price':2.00,'num':3,'sum':6.00,'type':'饮料'},{'name':'辣条','price':1.00,'num':5,'sum':5.00,'type':'食品'}],'time':'2022-04-18 12:13:14'}">> data.log
```

会看到如下输出:

```
……
-------------------------------------------
Time:1599885459000ms
-------------------------------------------
累计销售金额: 11.00
……
```

当我们使用 >> 往 /home/hadoop/data.log 文件中继续追加如下订单时:

```
[hadoop@master ~]$ echo "{'id':'102','info':[{'name':'洗衣机','price':2500.00,'num':1,'sum':2500.00,'type':'家电'}],'time':'2022-04-18 14:20:01'}">> data.log
```

会看到如下输出:

```
……
-------------------------------------------
Time:1598692164000 ms
-------------------------------------------
累计销售金额: 2511.00
……
```

针对需求二:统计当天某种类型的商品每个时间段内的销售金额,使用 ECharts 中的折线图,实时动态地显示。

(1)分析数据问题。

这里商品类型假设为食品,时间段为每小时,这样就需要分别统计一天内 24 个时间段的销售金额。因为数据是实时收集的,所以当前时间之后的时间段内是没有数据的。

(2)编写 Spark 程序。

由于订单数据是 JSON 格式,所以需要使用 JSON 解析包,这里使用阿里巴巴的 fastjson。计算结束后,还需要把结果存储到 Redis 中。

首先引入相关依赖。

```xml
<dependency>
    <groupId>com.alibaba</groupId>
    <artifactId>fastjson</artifactId>
    <version>1.2.73</version>
</dependency>
<dependency>
    <groupId>redis.clients</groupId>
    <artifactId>jedis</artifactId>
    <version>3.2.0</version>
</dependency>
```

然后编写 Spark 代码计算每个时间段内的销售金额，完整代码如程序清单 3.3 所示。

程序清单 3.3

```scala
package org.lanqiao.bigdata

import java.util.HashSet

import org.apache.kafka.clients.consumer.ConsumerConfig
import org.apache.kafka.common.serialization.StringDeserializer
import org.apache.spark.SparkConf
import org.apache.spark.streaming.Seconds
import org.apache.spark.streaming.StreamingContext
import org.apache.spark.streaming.kafka010.ConsumerStrategies
import org.apache.spark.streaming.kafka010.KafkaUtils
import org.apache.spark.streaming.kafka010.LocationStrategies

import com.alibaba.fastjson.JSON

import redis.clients.jedis.HostAndPort
import redis.clients.jedis.JedisCluster

object SparkOrder {
  def main(args:Array[String]):Unit={
valsparkconf=new SparkConf().setAppName("sparkorder")
    // 设置 DStream 批次时间间隔为 5 秒
valssc=new StreamingContext(sparkconf,Seconds(5))
    // 设置 Kafka 的配置：zk 集群链接、消费组
valkafkaParams=Map[String,Object](
    ConsumerConfig.BOOTSTRAP_SERVERS_CONFIG->"192.168.128.131:9092,
192.168.128.132:9092,192.168.128.133:9092",
```

```scala
        ConsumerConfig.GROUP_ID_CONFIG ->"mygroup",
        ConsumerConfig.KEY_DESERIALIZER_CLASS_CONFIG
        ->classOf[StringDeserializer],
        ConsumerConfig.VALUE_DESERIALIZER_CLASS_CONFIG
        ->classOf[StringDeserializer])
      // 设置消费主题
    val topics=Array("order")
      // 获取 Kafka 数据流
    val stream=KafkaUtils.createDirectStream[String,String](
    ssc,
    LocationStrategies.PreferConsistent,
    ConsumerStrategies.Subscribe[String,String](topics,kafkaParams))
      // 获取每一行的数据
    val line=stream.map(_.value)
      //val line=stream.map(_._2)
    line.print()
      // {'id':'101','info':[{'name':'可乐','price':2.00,'num':3,'sum':6.00,'type':'饮料'},
      //             {'name':'辣条','price':1.00,'num':5,'sum':5.00,'type':'食品'}],
      // 'time':'2022-04-18 12:13:14'}
      // 连接 Redis 集群
    valjedisClusterNodes=new HashSet[HostAndPort]
    jedisClusterNodes.add(new HostAndPort("192.168.128.131",6379))
    jedisClusterNodes.add(new HostAndPort("192.168.128.132",6379))
    jedisClusterNodes.add(new HostAndPort("192.168.128.133",6379))
    val jedis=new JedisCluster(jedisClusterNodes)
      // 解析数据
    val data=line.map {a=>{
      valjson=JSON.parseObject(a)
      valtime=json.getJSONString("time").substring(11,13)
      val info=json.getJSONArray("info")
      var sum=0.0
      for (i<-0 until info.size()){
          val type=info.getJSONObject(i).getString("type")
          if(type.equals("食品")){
              sum+=info.getJSONObject(i).getDouble("sum")
          }
      }
      // 分时间段累计求和并存入 Redis
      jedis.incrByFloat("shipin"+time,sum);
```

```
        // 求和后，把累计的返回值打印
        for(int i=0;i<24;i++){
            if(i<10){
                print(i+" 点 ~"+(i+1)+" 点累计销售金额: "+ jedis.get("shipin0"+i))
            }else{
                print(i+" 点 ~"+(i+1)+" 点累计销售金额: "+ jedis.get("shipin"+i))
            }
        }
}}

    // 开始计算
ssc.start()
    // 等待停止
ssc.awaitTermination()
    // 关闭 Jedis
jedis.close()
    }
}
```

程序清单 3.3 的大体思路如下。

① 连接 Kafka 服务器，每隔 5 秒尝试消费 order 主题下的订单消息，这里获取的是自上一次消费以来的 5 秒内的若干条订单数据。

② 对于每条订单数据，解析出食品类型和时间，根据时间属于哪个时间段，累加到对应的时间段上，并保存到 Redis 中。

③ 最后关闭相关资源。

（3）打包执行。

将程序打成 jar 包 SparkOrder.jar，上传到 Spark 集群，提交 Spark 任务。

```
[hadoop@master ~]$ spark-submit --master yarn --deploy-mode cluster \
--class org.lanqiao.bigdata.SparkOrder --executor-memory 1g \
/home/hadoop/SparkOrder.jar
```

启动 Spark Streaming 后，我们编写的程序就能获取 order 主题下的消息，解析 JSON 格式的消息并执行计算，将计算结果存储到 Redis 数据库中。

（4）验证数据结果。

我们可以到 Yarn 的 8088 端口上找到对应任务，然后去查看任务的 log 日志，程序中的输出将会体现在日志中。

当我们使用 >> 往 /home/hadoop/data.log 文件中追加如下订单时：

```
[hadoop@master ~]$ echo "{'id':'101','info':[{'name':'可乐','price':2.00,'num':3,'sum':6.00,'type':'饮料'},{'name':'辣条','price':1.00,'num':5,'sum':5.00,'type':'食品'}],'time':'2022-04-18 00:13:14'}">> data.log
[hadoop@master ~]$ echo "{'id':'102','info':[{'name':'可乐','price':2.00,'num':3,'sum':6.00,'type':'饮料'},{'name':'辣条','price':1.00,'num':5,'sum':5.00,'type':'食品'}],'time':'2022-04-18 01:13:14'}">> data.log
[hadoop@master ~]$ echo "{'id':'103','info':[{'name':'可乐','price':2.00,'num':3,'sum':6.00,'type':'饮料'},{'name':'辣条','price':1.00,'num':5,'sum':5.00,'type':'食品'}],'time':'2022-04-18 01:15:14'}">> data.log
[hadoop@master ~]$ echo "{'id':'104','info':[{'name':'可乐','price':2.00,'num':3,'sum':6.00,'type':'饮料'},{'name':'辣条','price':1.00,'num':5,'sum':5.00,'type':'食品'}],'time':'2022-04-18 02:13:14'}">> data.log
[hadoop@master ~]$ echo "{'id':'105','info':[{'name':'可乐','price':2.00,'num':3,'sum':6.00,'type':'饮料'},{'name':'辣条','price':1.00,'num':5,'sum':5.00,'type':'食品'}],'time':'2022-04-18 02:15:14'}">> data.log
[hadoop@master ~]$ echo "{'id':'106','info':[{'name':'可乐','price':2.00,'num':3,'sum':6.00,'type':'饮料'},{'name':'辣条','price':1.00,'num':5,'sum':5.00,'type':'食品'}],'time':'2022-04-18 02:23:14'}">> data.log
```

会看到如下输出：

```
……
-------------------------------------------
Time:1599886781000ms
-------------------------------------------
0 点~1 点累计销售金额：5.00
1 点~2 点累计销售金额：10.00
2 点~3 点累计销售金额：15.00
3 点~4 点累计销售金额：0.00
4 点~5 点累计销售金额：0.00
5 点~6 点累计销售金额：0.00
6 点~7 点累计销售金额：0.00
7 点~8 点累计销售金额：0.00
8 点~9 点累计销售金额：0.00
9 点~10 点累计销售金额：0.00
10 点~11 点累计销售金额：0.00
11 点~12 点累计销售金额：0.00
12 点~13 点累计销售金额：0.00
13 点~14 点累计销售金额：0.00
14 点~15 点累计销售金额：0.00
15 点~16 点累计销售金额：0.00
16 点~17 点累计销售金额：0.00
```

```
17 点~18 点累计销售金额：0.00
18 点~19 点累计销售金额：0.00
19 点~20 点累计销售金额：0.00
20 点~21 点累计销售金额：0.00
21 点~22 点累计销售金额：0.00
22 点~23 点累计销售金额：0.00
23 点~24 点累计销售金额：0.00
……
```

可以看到，0 点到 1 点之间一共有 1 笔订单，累计销售金额为 5.00；1 点到 2 点之间一共有 2 笔订单，累计销售金额为 5.00+5.00=10.00；2 点到 3 点之间一共有 3 笔订单，累计销售金额为 5.00+5.00+5.00=15.00；其余时间暂时没有订单成交或者还未到来。

当我们使用 >> 往 /home/hadoop/data.log 文件中继续追加如下订单时：

```
[hadoop@master ~]$ echo "{'id':'107','info':[{'name':'可乐','price':2.00,'num':3,'sum':6.00,'type':'饮料'},{'name':'辣条','price':1.00,'num':2,'sum':2.00,'type':'食品'}],'time':'2022-04-18 03:20:01'}">> data.log
[hadoop@master ~]$ echo "{'id':'108','info':[{'name':'可乐','price':2.00,'num':3,'sum':6.00,'type':'饮料'},{'name':'辣条','price':1.00,'num':4,'sum':4.00,'type':'食品'}],'time':'2022-04-18 03:53:01'}">> data.log
[hadoop@master ~]$ echo "{'id':'109','info':[{'name':'可乐','price':2.00,'num':3,'sum':6.00,'type':'饮料'},{'name':'辣条','price':1.00,'num':3,'sum':3.00,'type':'食品'}],'time':'2022-04-18 04:21:01'}">> data.log
[hadoop@master ~]$ echo "{'id':'110','info':[{'name':'可乐','price':2.00,'num':3,'sum':6.00,'type':'饮料'},{'name':'辣条','price':1.00,'num':5,'sum':5.00,'type':'食品'}],'time':'2022-04-18 04:32:01'}">> data.log
[hadoop@master ~]$ echo "{'id':'111','info':[{'name':'可乐','price':2.00,'num':3,'sum':6.00,'type':'饮料'},{'name':'辣条','price':1.00,'num':8,'sum':8.00,'type':'食品'}],'time':'2022-04-18 04:43:01'}">> data.log
[hadoop@master ~]$ echo "{'id':'112','info':[{'name':'可乐','price':2.00,'num':3,'sum':6.00,'type':'饮料'},{'name':'辣条','price':1.00,'num':4,'sum':4.00,'type':'食品'}],'time':'2022-04-18 05:20:01'}">> data.log
```

会看到如下输出：

```
……
-------------------------------------------
Time:1598699825000 ms
-------------------------------------------
0 点~1 点累计销售金额：5.00
1 点~2 点累计销售金额：10.00
2 点~3 点累计销售金额：15.00
```

```
3 点 ~4 点累计销售金额: 6.00
4 点 ~5 点累计销售金额: 16.00
5 点 ~6 点累计销售金额: 4.00
6 点 ~7 点累计销售金额: 0.00
7 点 ~8 点累计销售金额: 0.00
8 点 ~9 点累计销售金额: 0.00
9 点 ~10 点累计销售金额: 0.00
10 点 ~11 点累计销售金额: 0.00
11 点 ~12 点累计销售金额: 0.00
12 点 ~13 点累计销售金额: 0.00
13 点 ~14 点累计销售金额: 0.00
14 点 ~15 点累计销售金额: 0.00
15 点 ~16 点累计销售金额: 0.00
16 点 ~17 点累计销售金额: 0.00
17 点 ~18 点累计销售金额: 0.00
18 点 ~19 点累计销售金额: 0.00
19 点 ~20 点累计销售金额: 0.00
20 点 ~21 点累计销售金额: 0.00
21 点 ~22 点累计销售金额: 0.00
22 点 ~23 点累计销售金额: 0.00
23 点 ~24 点累计销售金额: 0.00
……
```

可以看到，3 点到 4 点之间一共有 2 笔订单，累计销售金额为 2.00+4.00=6.00；4 点到 5 点之间一共有 3 笔订单，累计销售金额为 3.00+5.00+8.00=16.00；5 点到 6 点之间一共有 1 笔订单，累计销售金额为 4.00，其余时间暂时没有订单成交或者还未到来。

3.5 任务五：Java+ECharts 数据可视化

Java+ECharts 数据可视化的流程如图 3.5 所示。

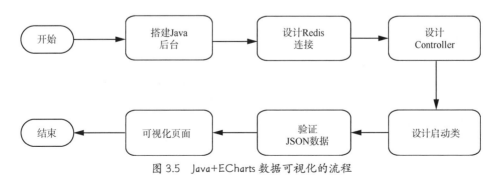

图 3.5　Java+ECharts 数据可视化的流程

按照操作步骤，可以将 Java+ECharts 数据可视化的过程划分为 6 个阶段，分别是搭建 Java 后台、设计 Redis 连接、设计 Controller、设计启动类、验证 JSON 数据和可视化页面。

针对需求一：统计当天的累计销售金额，使用 ECharts 中的折线图，实时动态地显示。

（1）搭建 Java Web 后台开发框架（SpringBoot）。

创建 SpringBoot 项目，步骤如下。

① 打开 Eclipse，新建 Spring Boot → Spring Starter Project，单击"next"。

② 选择 jar 方式，设置 java 为"java8"，group 为"org.lanqiao"，Artifact 为"bigdata"，Package 为"org.lanqiao.bigdata"，单击"next"。

③ 选择事先要加载的组件，即 nosql 里的 Spring Data Redis（Access+Driver），Web 里的 Spring Web，单击"next"和"finish"。

④ 单击操作完成后，需要稍等一会儿，maven 会自动下载相关 jar 包资源。

（2）设计 Redis 连接。

打开 src/main/resources 目录下的 application.properties 文件，内容如程序清单 3.4 所示。

程序清单 3.4

```
# REDIS (RedisProperties)
# Redis 数据库索引（默认为 0）
spring.redis.database=0
# Redis 服务器地址
spring.redis.host=localhost
# Redis 服务器连接端口
spring.redis.port=6379
# Redis 服务器连接密码（默认为空）
spring.redis.password=
# 连接池最大连接数（使用负值表示没有限制）
spring.redis.pool.max-active=8
# 连接池最大阻塞等待时间（使用负值表示没有限制）
spring.redis.pool.max-wait=-1
# 连接池中的最大空闲连接
spring.redis.pool.max-idle=8
# 连接池中的最小空闲连接
spring.redis.pool.min-idle=0
# 连接超时时间（毫秒）
spring.redis.timeout=0
```

（3）设计 Controller。

创建包 org.lanqiao.bigdata.controller，然后在该包下创建类 A1Controller，代码如程序清单 3.5 所示。

程序清单 3.5
```java
package org.lanqiao.bigdata.controller;

import java.util.HashMap;
import java.util.Map;
import org.springframework.beans.factory.annotation.Autowired;
import org.springframework.data.redis.core.StringRedisTemplate;
import org.springframework.web.bind.annotation.RequestMapping;
import org.springframework.web.bind.annotation.RestController;

@RestController
public class A1Controller {
@Autowired
private static StringRedisTemplatestringRedisTemplate;

    @RequestMapping("/a1")
    public Map<String,Double> getA1(){
        String value=stringRedisTemplate.opsForValue().get("orderSum");
        Map map=new HashMap<String,Double>();
        map.put("y",Double.parseDouble(value));
        return map;
    }
}
```

（4）设计启动类。

在包 org.lanqiao.bigdata 下创建启动类 DemoApplication，代码如程序清单 3.6 所示。

程序清单 3.6
```java
package org.lanqiao.bigdata;

import org.springframework.boot.SpringApplication;
import org.springframework.boot.autoconfigure.SpringBootApplication;
import org.springframework.context.annotation.ComponentScan;

@SpringBootApplication
@ComponentScan(basePackages={"org.lanqiao.bigdata"})
public class DemoApplication {
```

```
    public static void main(String[] args){
SpringApplication.run(DemoApplication.class,args);
    }
}
```

（5）验证 JSON 数据。

右击项目 → 选择"run as" → 选择"spring boot app"，打开浏览器，输入地址"http://localhost:8080/a1"，查看 JSON 数据，具体数值要看已经采集的订单情况。

假设目前已经采集了 1 笔订单：

{'id':'101','info':[{'name':'可乐','price':2.00,'num':3,'sum':6.00,'type':'饮料'},{'name':'辣条','price':1.00,'num':5,'sum':5.00,'type':'食品'}],'time':'2022-04-18 12:13:14'}

那么此时的数值应该为

{"y":11.00}

假设目前已经采集了 2 笔订单：

{'id':'101','info':[{'name':'可乐','price':2.00,'num':3,'sum':6.00,'type':'饮料'},{'name':'辣条','price':1.00,'num':5,'sum':5.00,'type':'食品'}],'time':'2022-04-18 12:13:14'}

{'id':'102','info':[{'name':'洗衣机','price':2500.00,'num':1,'sum':2500.00,'type':'家电'}],'time':'2022-04-18 14:20:01'}

那么此时的数值应该为

{"y":2511.00}

（6）可视化页面。

这里我们选择 ECharts 中的折线图，并且实时动态地滚动数据，x 轴为时间，y 轴为数值，每隔 5 秒钟自动从右往左滚动。

在项目 static 目录下创建 a1.html，代码如程序清单 3.7 所示。

程序清单 3.7

```
<!DOCTYPE html>
<html>
<head>
    <meta charset="utf-8">
    <title>ECharts</title>
    <!-- 引入 echarts.js -->
    <script src="echarts.js"></script>
    <!-- 引入 jquery-->
    <script src="jquery-1.8.3.min.js"></script>
```

```html
</head>
<body>
    <!-- 为ECharts准备一个具备大小（宽高）的图表显示区域 -->
    <div id="main" style="width:1400px;height:600px;"></div>
    <script type="text/javascript">
    var now=new Date();
    var x=[];//x轴数据
    var y=[];//y轴数据
    var myChart=echarts.init(document.getElementById('main'));
    setInterval(function () {
            $.get('a1').done(function(data) {
                // 超过2分钟时删除最早的一个数据
                if(data.length>24){
                    x.shift();
                    y.shift();
                }
                // 添加最新的一个数据
                x.push(now.getHours()+":"+now.getMinutes()+":"+now.getSeconds());
                y.push(data.y);
                // 重新设置option
                myChart.setOption({
                    // 标题
                    title:{
                        text:'累计销售金额',
                        subtext:'每5秒钟更新一次最近2分钟内的数据'
                    },
                    // 图例
                    legend:{
                        data:['累计销售金额']
                    },
                    // 提示框内容由坐标轴触发
                    tooltip:{
                        trigger:'axis',
                        formatter:function(params){
                            params=params[0];
                            var time=params.name.split(":");
                            //10:10:10
                            return
now.getFullYear()+'/'+(now.getMonth()+1)+'/'+now.getDate()+' '+time[0]+':'+time[1
]+":"+time[2] +'-'+ params.value+'元';
```

```
                },
                axisPointer:{
                    animation:false
                }
            },
            //x轴内容
            xAxis:{
                type:'category',
                splitLine:{
                    show:false
                },
                axisLabel:{// 文字显示
                    interval:0,// 间隔 0 表示每一个都强制显示
                    rotate:40// 旋转角度 [-90,90]
                },
                data:x
            },
            //y轴内容
            yAxis:{
                type:'value',
                boundaryGap:[0,'100%'],
                splitLine:{
                    show:false
                }
            },
            // 绘图类型和数据
            series:[{
                name:'累计销售金额',
                type:'line',
                showSymbol:false,
                hoverAnimation:false,
                data:y
            }]
        });
    })
},5000);
        </script>
    </body>
</html>
```

其中，使用 setInterval 方法设置每隔 5000 毫秒，异步请求一次数据后，通过 setOption() 方法重置 option 来刷新图表，这样就会呈现实时的数据走势。一旦数据的个数超过 24 个

（2分钟），就删除数据数组最开始的一个数据，这样图形始终呈现最近的24个数据，页面内容不会过多，恰到好处。x轴内容由JavaScript动态生成当前时间，y轴内容由后端请求处理后返回。

代码编写完毕，项目启动并运行后，打开浏览器，输入地址"http://localhost:8080/a1.html"，然后准备一批订单数据，使用 >> 往 /home/hadoop/data.log 文件中追加订单，就能看到可视化数据了，如图3.6所示。至于订单的金额，自己设定即可。

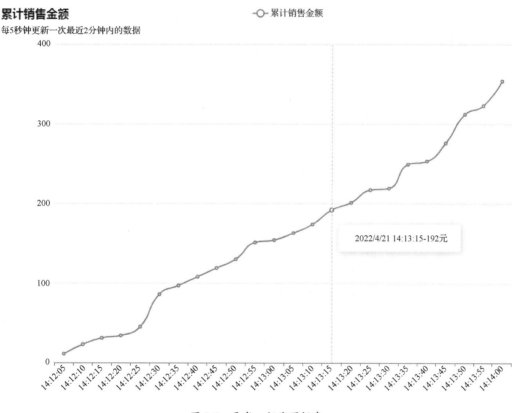

图3.6 需求一折线图报表

针对需求二：统计当天某种类型的商品每个时间段内的销售金额，使用ECharts中的折线图，实时动态地显示。

（1）搭建Java Web后台开发框架（SpringBoot）。

项目后台之前已经搭建完毕。

（2）设计 Redis 连接。

Redis 连接参数保持不变。

（3）设计 Controller。

在包 org.lanqiao.bigdata.controller 下创建类 A2Controller，代码如程序清单 3.8 所示。

程序清单 3.8

```java
package org.lanqiao.bigdata.controller;

import java.util.HashMap;
import java.util.Map;
import org.springframework.beans.factory.annotation.Autowired;
import org.springframework.data.redis.core.StringRedisTemplate;
import org.springframework.web.bind.annotation.RequestMapping;
import org.springframework.web.bind.annotation.RestController;

@RestController
public class A2Controller {
@Autowired
private static StringRedisTemplatestringRedisTemplate;

    @RequestMapping("/a2")
    public Map<String,Object>getA2(){
        Double[] y=new Double[24];
        for(int i=0;i<24;i++){
            if(i<10){
                y[i]=Double.parseDouble(stringRedisTemplate.opsForValue().get("shipin0"+i));
            }else{
                y[i]=Double.parseDouble(stringRedisTemplate.opsForValue().get("shipin"+i));
            }
        }
        Map map=new HashMap<String,Object>();
        map.put("y",y);
        return map;
    }
}
```

（4）设计启动类。

启动类保持不变。

（5）验证 JSON 数据。

右击项目 → 选择 "run as" → 选择 "spring boot app"，打开浏览器，输入地址 "http://localhost:8080/a2"，查看 JSON 数据，具体数值要看已经采集的订单情况。

假设目前已经采集了如下订单：

```
[hadoop@master ~]$ echo "{'id':'101','info':[{'name':'可乐','price':2.00,'num':3,'sum':6.00,'type':'饮料'},{'name':'辣条','price':1.00,'num':5,'sum':5.00,'type':'食品'}],'time':'2022-04-18 00:13:14'}">> data.log
[hadoop@master ~]$ echo "{'id':'102','info':[{'name':'可乐','price':2.00,'num':3,'sum':6.00,'type':'饮料'},{'name':'辣条','price':1.00,'num':5,'sum':5.00,'type':'食品'}],'time':'2022-04-18 01:13:14'}">> data.log
[hadoop@master ~]$ echo "{'id':'103','info':[{'name':'可乐','price':2.00,'num':3,'sum':6.00,'type':'饮料'},{'name':'辣条','price':1.00,'num':5,'sum':5.00,'type':'食品'}],'time':'2022-04-18 01:15:14'}">> data.log
[hadoop@master ~]$ echo "{'id':'104','info':[{'name':'可乐','price':2.00,'num':3,'sum':6.00,'type':'饮料'},{'name':'辣条','price':1.00,'num':5,'sum':5.00,'type':'食品'}],'time':'2022-04-18 02:13:14'}">> data.log
[hadoop@master ~]$ echo "{'id':'105','info':[{'name':'可乐','price':2.00,'num':3,'sum':6.00,'type':'饮料'},{'name':'辣条','price':1.00,'num':5,'sum':5.00,'type':'食品'}],'time':'2022-04-18 02:15:14'}">> data.log
[hadoop@master ~]$ echo "{'id':'106','info':[{'name':'可乐','price':2.00,'num':3,'sum':6.00,'type':'饮料'},{'name':'辣条','price':1.00,'num':5,'sum':5.00,'type':'食品'}],'time':'2022-04-18 02:23:14'}">> data.log
```

那么此时的 JSON 数值应该为

```
{"y":5,10,15,0,0,0,0,0,0,0,0,0,0,0,0,0,0,0,0,0,0,0,0,0}
```

又继续采集了如下订单：

```
[hadoop@master ~]$ echo "{'id':'107','info':[{'name':'可乐','price':2.00,'num':3,'sum':6.00,'type':'饮料'},{'name':'辣条','price':1.00,'num':2,'sum':2.00,'type':'食品'}],'time':'2022-04-18 03:20:01'}">> data.log
[hadoop@master ~]$ echo "{'id':'108','info':[{'name':'可乐','price':2.00,'num':3,'sum':6.00,'type':'饮料'},{'name':'辣条','price':1.00,'num':4,'sum':4.00,'type':'食品'}],'time':'2022-04-18 03:53:01'}">> data.log
[hadoop@master ~]$ echo "{'id':'109','info':[{'name':'可乐','price':2.00,'num':3,'sum':6.00,'type':'饮料'},{'name':'辣条','price':1.00,'num':3,'sum':3.00,'type':'食品'}],'time':'2022-04-18 04:21:01'}">> data.log
[hadoop@master ~]$ echo "{'id':'110','info':[{'name':'可乐','price':2.00,'num':3,'sum':6.00,'type':'饮料'},{'name':'辣条','price':1.00,'num':5,'sum':5.00,'type':'食品'}],'time':'2022-04-18 04:32:01'}">> data.log
```

```
[hadoop@master ~]$ echo "{'id':'111','info':[{'name':'可乐','price':2.00,'num':
3,'sum':6.00,'type':'饮料'},{'name':'辣条','price':1.00,'num':8,'sum':8.00,'type':
'食品'}],'time':'2022-04-18 04:43:01'}" >> data.log
[hadoop@master ~]$ echo "{'id':'112','info':[{'name':'可乐','price':2.00,'num':
3,'sum':6.00,'type':'饮料'},{'name':'辣条','price':1.00,'num':4,'sum':4.00,'type':
'食品'}],'time':'2022-04-18 05:20:01'}" >> data.log
```

那么此时的 JSON 数值应该为

```
{"y":5,10,15,6,16,4,0,0,0,0,0,0,0,0,0,0,0,0,0,0,0,0,0,0}
```

（6）可视化页面。

这里我们选择 ECharts 中的折线图，x 轴为 24 个小时，y 轴为销售金额，每隔 5 秒钟取一次值更新数据。

在项目 static 目录下创建 a2.html，代码如程序清单 3.9 所示。

程序清单 3.9

```html
<!DOCTYPE html>
<html>
<head>
    <meta charset="utf-8">
    <title>ECharts</title>
    <!-- 引入 echarts.js -->
    <script src="echarts.js"></script>
    <!-- 引入 jquery-->
    <script src="jquery-1.8.3.min.js"></script>
</head>
<body>
    <!-- 为ECharts准备一个具备大小（宽高）的图表显示区域 -->
    <div id="main" style="width:1400px;height:600px;"></div>
    <script type="text/javascript">
    var now=new Date();
    var myChart=echarts.init(document.getElementById('main'));
    setInterval(function () {
        $.get('a2').done(function(data){
            // 重新设置 option
            myChart.setOption({
                // 标题
                title:{
                    text:'食品每小时销售金额分布图',
                    subtext:'每 5 秒钟更新一次数据'
                },
```

```
                    //图例
                    legend:{
                        data:['每小时销售金额']
                    },
                    //提示框内容由坐标轴触发
                    tooltip:{
                        trigger:'axis',
                        formatter:function (params) {
                            params=params[0];
                            var time=params.name;
                            return
now.getFullYear()+'/'+(now.getMonth()+1)+'/'+now.getDate()+' '+time+' 点 -'+
params.value+'元';
                        },
                        axisPointer:{
                            animation:false
                        }
                    },
                    //x轴内容
                    xAxis:{
                        name:'点',
                        type:'category',
                        splitLine:{
                            show:false
                        },
                        axisLabel:{//文字显示
                            interval:0,//间隔0表示每一个都强制显示
                            rotate:40//旋转角度[-90,90]
                        },
data:[ "00","01","02","03","04","05","06","07","08","09","10","11","12",
"13","14","15","16","17","18","19","20","21","22","23"]
                    },
                    //y轴内容
                    yAxis:{
                        type:'value',
                        boundaryGap:[0,'100%'],
                        splitLine:{
                            show:false
                        }
                    },
                    //绘图类型和数据
```

```
                    series:[{
                        name:'每小时销售金额',
                        type:'line',
                        showSymbol:false,
                        hoverAnimation:false,
                        data:data.y
                    }]
                });
            })
        },5000);
        </script>
    </body>
</html>
```

其中，使用 setInterval 方法设置每隔 5000 毫秒，异步请求一次数据后，通过 setOption() 方法重置 option 来刷新图表，这样就会呈现实时的数据走势。

代码编写完毕，项目启动并运行后，打开浏览器，输入地址"http://localhost:8080/a2.html"，然后准备一批订单数据，使用 >> 往 /home/hadoop/data.log 文件中追加订单，然后就能看到可视化数据了，如图 3.7 所示。

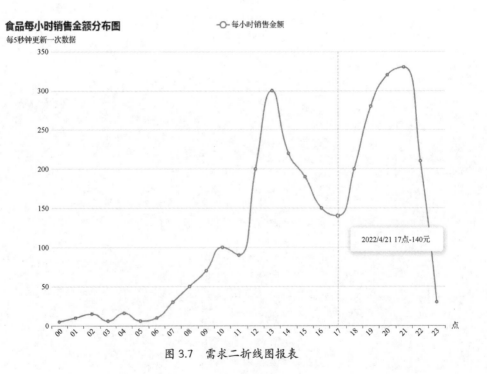

图 3.7　需求二折线图报表

3.6 答疑解惑

问题1：项目中使用 Spark 而没有使用 MapReduce 来计算，前者有什么优势吗？

解答1：Spark 是借鉴了 MapReduce，并在其基础上发展起来的，继承了其分布式计算的优点并进行了改进，相对于 MapReduce 而言具有如下的优势。

（1）Spark 把运算的中间数据（Shuffle 阶段产生的数据）存放在内存，迭代计算效率更高，而 MapReduce 的中间结果需要落地，保存到磁盘中。

（2）Spark 容错性高，它通过弹性分布式数据集 RDD 来实现高效容错。RDD 是一组分布式的存储在节点内存中的只读性的数据集，这些集合是弹性的，某一部分丢失或者出错，可以通过整个数据集的计算流程的血缘关系来实现重建。而用 MapReduce，只能重新计算。

（3）Spark 更通用，提供了转换和动作这两大类的功能，另外还有流式处理模块、图计算等，而 MapReduce 只提供了 map 和 reduce 两种操作，流计算及其他的模块支持比较缺乏。

（4）Spark 框架和生态更为复杂，有 RDD，有血缘，有执行时的有向无环图（DAG），有 stage 划分等，很多时候 Spark 作业都需要根据不同业务场景的需要进行调优以达到性能要求。而 MapReduce 框架及其生态相对较为简单，对性能的要求也相对较低。

问题2：Spark 在项目中使用的是 Standalone 模式，还有其他模式吗？

解答2：Spark 支持 3 种集群管理器，分别如下。

（1）Standalone 模式。资源管理器是 master 节点，调度策略相对单一，只支持先进先出模式。

（2）Hadoop Yarn 模式。资源管理器是 Yarn 集群，主要用来管理资源。Yarn 支持动态资源的管理，还可以调度其他实现了 Yarn 调度接口的集群计算，非常适用于多个集群同时部署的场景，是目前最流行的一种资源管理系统。

（3）Apache Mesos 模式。Mesos 是专门用于分布式系统资源管理的开源系统，与 Yarn 一样用 C++ 开发，可以对集群中的资源做弹性管理。

问题3：Spark RDD 中 map 与 flatMap 有什么区别？

解答3：map 对 RDD 的每个元素进行转换，文件中的每一行数据返回一个数组对象。

flatMap 对 RDD 的每个元素进行转换，然后扁平化。将所有的对象合并为一个对象，文件中的所有行数据仅返回一个数组对象，会抛弃值为 null 的值。

问题 4：Spark RDD 如何区分窄依赖和宽依赖？

解答 4：窄依赖和宽依赖的主要区别如下。

（1）窄依赖指的是每一个父 RDD 的 Partition（分区）最多被子 RDD 的一个 Partition 使用。

（2）宽依赖指的是多个子 RDD 的 Partition 会依赖同一个父 RDD 的 Partition。

问题 5：Kafka 中的消息在消费时，能做到有序吗？

解答 5：在 Kafka 中，Partition 是真正保存消息的地方，发送的消息都存放在这里。Partition 又存在于 Topic（主题）中，并且一个 Topic 可以指定多个 Partition。

在 Kafka 中，只保证每个 Partition 内有序，不保证 Topic 所有分区都是有序的。

可通过以下两种方法来保证 Kafka 消息的消费顺序。

（1）1 个 Topic 只创建 1 个 Partition，这样生产者的所有数据都发送到了一个 Partition 内，保证了消息的消费顺序。

（2）生产者在发送消息的时候指定要发送到哪个 Partition，这样可以保证消息的局部有序。

问题 6：Spark RDD 通过什么方式容错？

解答 6：Spark RDD 采用记录更新的方式来容错。一般来说，如果更新粒度太细，那么记录更新成本也不低。因此，RDD 只支持粗粒度转换，即只记录单个块上执行的单个操作，然后将创建 RDD 的一系列变换序列（每个 RDD 都包含了它是如何由其他 RDD 变换过来的以及如何重建某一块数据的信息的，因此 RDD 的容错机制又称"血统容错"）记录下来，以便恢复丢失的分区。lineage（血统）本质上与数据库中的重做日志（Redo Log）很类似，只不过重做日志粒度很大，是通过重做全局数据进而恢复数据的。

3.7 拓展练习

1. 拓展项目

我们拓展一下本章项目，假设企业高层希望能够随时查看当天的累计订单数量，这样可以和往日进行对比。于是我们提出如下需求：实时地统计当天的累计订单数量，使用 ECharts 中的折线图，动态地显示。按照本章所讲的步骤来做，请读者自行完成该任务。

2. 强化 Spark RDD 编程

假设有如下文件 a.txt：

```
Tom,english,80
Tom,maths,50
Tom,chinese,60
Jerry,english,90
Jerry,maths,60
Jerry,chinese,80
LiLei,english,70
LiLei,maths,80
```

其中第一列是姓名，第二列是科目，第三列是成绩。

请使用 Spark RDD 来回答以下问题：

（1）一共有多少名学生？

（2）Tom 的总分是多少？

（3）科目 english 的平均分是多少？

（4）每名同学选修的课程数是多少？

第4章
金融大数据分析项目

项目标签：职业院校技能大赛（Spark、Flink）

【本章简介】

本章通过介绍高职院校职业技能大赛大数据技术与应用赛项，帮助读者熟悉这一赛项的考核情况，并掌握大数据相关的职业技能，提高大数据的处理能力。相对于第2章的项目，本项目的内容更加丰富，技术要求也更加灵活多样，既有离线处理，又有实时处理。

【知识地图】

本章项目涉及的知识地图如图4.1所示，读者在学习过程中可以查询相关知识卡片，补充理论知识。

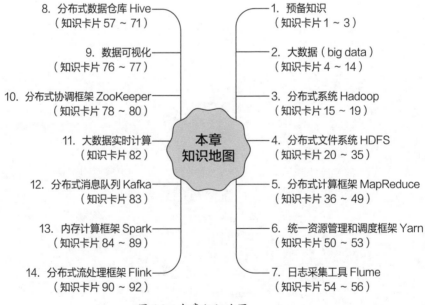

图4.1 本章知识地图

4.1 大赛简介

1．竞赛目的

为适应大数据产业对高素质技能型人才的职业需求，××省开展以大数据技术与应用为核心内容和工作基础的竞赛活动。该赛项重点考查参赛选手基于 Spark、Flink 平台环境，充分利用 Spark Core、Spark SQL、Flume、Kafka、Flink 等相关技术的特点，基于 Scala、JavaScript 等开发语言，综合软件开发相关技术，解决实际问题的能力，力图激发学生对大数据相关知识和技术的学习兴趣，提升学生职业素养和职业技能，从而努力为中国大数据产业的发展储备及输送新鲜血液。

通过举办本赛项，可以搭建校企合作的平台，提升大数据技术与应用专业及其他相关专业毕业生的能力和素质，满足企业用人需求，促进校企合作协同育人，对接产业发展，实现行业资源、企业资源与教学资源的有机融合，促使高职院校在专业建设、课程建设、人才培养方案和人才培养模式等方面，跟踪社会发展的最新需要，缩小人才培养与行业需求差距，引领职业院校专业建设与课程改革。

2．竞赛内容

赛项以大数据技术与应用为核心内容，重点考查参赛选手基于 Spark、Flink 平台环境，充分利用 Spark Core、Spark SQL、Flume、Kafka、Flink 等技术的特点，综合软件开发相关技术，解决实际问题的能力，具体包括以下要点。

（1）掌握基于 Spark 的离线分析平台、基于 Flink 的实时分析平台，按照项目需求安装相关技术组件并进行合理配置。

（2）掌握基于 Spark 的离线数据抽取相关技术，完成指定数据的抽取并写入 Hive 表中。

（3）综合利用 Spark Core、Spark SQL 等技术，使用 Scala 开发语言，完成某金融大数据分析项目的离线统计服务，并将统计结果存入 MySQL 数据库中。

（4）综合利用 Flume、Kafka、Flink 等相关技术，使用 Scala 开发语言，完成某金融大数据分析项目的实时计算需求，并将计算结果存入 Redis 和 MySQL 数据库中。

（5）综合运用 HTML、CSS、JavaScript 等开发语言，结合 Vue.js 前端技术、ECharts 数据可视化组件，对 MySQL 中的数据进行可视化呈现。

（6）利用 Scala 开发语言，基于 Spark ML 机器学习库，根据既有数据建立数据模型完成数据分析、数据挖掘操作。

（7）根据数据可视化结果，完成数据分析报告的编写。

3．竞赛技术平台

（1）硬件环境。

竞赛硬件设备如表 4.1 所示。

表 4.1　竞赛硬件设备

设备类别	数量	设备用途	基本配置
竞赛服务器	每支参赛队伍 1 台。根据参赛队数量，配备 10% 的备份机器	构建大数据平台集群	性能相当于 i5 处理器，32GB 以上内存，1TB 以上硬盘，配备网卡（千兆），显示器分辨率要求在 1024 像素 ×768 像素以上
竞赛客户机	每支参赛队伍 3 台。根据参赛团队数量，配备 10% 的备份机器	竞赛选手比赛使用	性能相当于 i5 处理器，8GB 以上内存，1TB 以上硬盘，显示器分辨率要求在 1024 像素 ×768 像素以上

其中竞赛服务器主要功能如下。

① 提供镜像上传存储、云主机一键创建、计算存储规格创建及清理、云主机回收站、云主机在线克隆为镜像模板、云主机配置参数修改等功能。

② 支持多账户管理、云主机在线或远程访问等功能；支持多角色如（管理员、教师、学生）管理、专业管理、班级管理、用户管理、操作日志、系统设置、环境池管理、弹性 IP 管理、镜像管理、实训模板管理、课程管理、实训监控、实验环境、在线实训等功能；支持通过 VNC、SSH 等多种模式访问竞赛平台。

③ 支持模拟大数据平台环境搭建、离线数据抽取、离线数据统计、实时数据采集与实时计算、数据可视化等贯穿大数据技术并符合产业主流需求的技术技能应用，提供大数据竞赛所需的虚拟机，所涉及的开发语言包括 Scala、JavaScript、HTML 等。

（2）软件环境。

竞赛软件环境如表 4.2 所示。

表 4.2　竞赛软件环境

设备类型	软件类别	软件名称、版本号
竞赛服务器	大数据集群操作系统	CentOS 7
	大数据平台组件	Hadoop 2.7.7
		Hive 2.3.4
		Spark 2.1.1
		Kafka 2.0.0
		Redis 4.0.1
		Flume 1.7.0
		Flink 1.10.2
		JDK 1.8
		MySQL 5.7
开发客户端	PC 操作系统	Ubuntu 18.04（64 位）
	浏览器	Chrome
	开发语言	Scala 2.11
	开发工具	IDEA 2019(Community Edition)
		Visual Studio Code 1.58
	数据可视化组件	Vue.js 3.0
		ECharts 5.1
	文档编辑器	WPS Linux 版
	输入法	搜狗拼音输入法 Linux 版

4.2　任务一：需求分析

1. 业务背景

银行业一直是一个数据驱动的行业，数据也一直是银行信息化的主题词。银行的信息化进程先后经历过业务电子化、数据集中化、管理模型化等阶段，如今随着大数据技术的飞速发展，银行信息化也进入了新的阶段——大数据时代。

目前，国内的银行都积累了海量的金融数据，包括各类结构化、半结构化、非结构化数据，存储方式多样。比如银行拥有客户姓名、年龄、性别、婚姻状况、职业、账龄、储蓄存款账户、是否使用信用卡、平均余额、信用交易数量、借记交易数量等数据信息。但是这些海量数据还没得到充分利用，显得价值含量较低。只有经过合适的预处理、模型设计、分析挖掘后，才能发现隐藏在其中的规律。而应用大数据分析技术，可以从海量的、有噪声的、模糊的、随机的数据中提取隐含在其中的、人们事先不知道的、但又有用的信息和知识。银行可以利用这些信息和知识来提升金融业务的服务效率和管理水平，银行的关键业务也能从中获得巨大收益。

银行在大数据技术应用方面具有天然优势：一方面，银行在业务开展过程中积累了大量有价值的数据，这些数据经过大数据技术的挖掘和分析之后，将产生巨大的商业价值；另一方面，银行在资金、设备、人才、技术上都具有极大的便利条件，有能力采用大数据的最新技术。通过建立大数据分析平台，可以对金融数据进行挖掘、分析，创造数据增值价值，提供精准营销、业务体验优化、客户综合管理、风险控制等多种金融服务。

2．业务需求

（1）银行高层希望了解银行有哪些业务办理的方式？各自占多少份额？这样可以优化现有的资源配置，更好地服务客户。

（2）银行高层希望了解在银行所有业务中，各个城市的业务量分别是多少？这样有利于开展线下业务活动。

（3）银行高层想知道今年以来各个月份的业务量及变化趋势，并预测今年的业务办理总量。

（4）银行高层希望能够给基金用户推荐合适的基金。

3．功能需求

（1）统计分析银行各种业务办理方式的占比，并采用 ECharts 饼图可视化。

（2）统计分析银行各个城市的业务量，并采用 ECharts 柱状图可视化。

（3）统计分析银行截止到目前，每个月的业务量，并采用 ECharts 折线图可视化。

（4）通过历史用户购买基金的数据，使用协同过滤算法，给当前用户推荐合适的基金。

其中（1）和（2）属于离线数据分析，（3）属于实时数据分析，（4）属于数据挖掘。

4.3 任务二：项目流程

1．技术领域判定

从需求分析中，我们知道该项目属于综合型项目，涵盖了离线数据分析和实时数据分析，具有以下几个特征。

（1）数据量较大。

（2）业务较为复杂。

（3）需要有简单的报表。

（4）技术覆盖了大部分大数据框架。

这是一个综合型金融大数据分析项目，部分数据处理需求对实时性要求不高，相关业务指标的统计分析用到了离线处理的相关技术，如统计分析银行所有业务中各种业务办理方式的占比。而其他需求则要求具有一定的实时性，相关业务指标的计算用到了实时处理的相关技术，如统计分析截止到目前，银行每个月的业务量。

2．技术架构分析

该项目的大数据架构由下往上可以分为 4 层：数据采集层、数据处理层、数据存储层、数据展示层。下面我们对这 4 层做简单分析。

（1）数据采集层。离线处理方向使用 Spark SQL 从 MySQL 中采集数据；实时处理方向使用 Flume 和 Kafka 从服务器采集数据。

（2）数据处理层。离线处理方向使用 Spark Core 处理数据；实时处理方向使用 Flink 处理数据。

（3）数据存储层。离线处理方向使用 MySQL 存储数据；实时处理方向使用 Redis 存储数据。

（4）数据展示层。离线处理方向和实时处理方向都使用 Spring Boot 结合 Vue.js、ECharts 等前端技术，以图形化的方式来展示数据结果，如饼图、柱状图、折线图等。

3．实现流程

本项目的流程，如图 4.2 所示。数据处理分为 2 个方向，一个是离线处理方向，另一个是实时处理方向。离线处理方向首先使用 Spark SQL 从 MySQL 数据库中抽取数据，并存储到 Hive 分区表中；然后使用 Spark Core 对数据进行数据清洗和统计分析，并把数据分析结果保存到 MySQL 中；最后创建 Spring Boot 项目，结合 Vue.js、ECharts 实现数据可视化。实时处理方向首先使用 Flume 采集服务器上的数据,并存储到 Kafka 消息中间件中;然后使用 Flink 从 Kafka 中提取消息进行实时数据计算,再把计算结果存储到 Redis 数据库中;最后创建 Spring Boot 应用程序来访问 Redis 数据库,使用 Vue.js 和 ECharts 呈现在网页上。

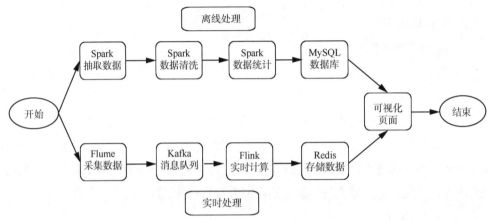

图 4.2　金融大数据分析项目的流程

4．开发环境准备

在完成后续任务之前,我们需要安装部署并配置一系列的软件,准备好开发环境,如表 4.3 所示。这些基础操作,我们放入附录中以供不熟悉的同学参考。

表 4.3　开发环境准备

操作	附录
Hadoop 安装部署和配置	附录 1
Flume 安装部署和配置	附录 5
Hive 安装部署和配置	附录 6
Sqoop 安装部署和配置	附录 7
Hadoop 高可用集群环境安装部署和配置	附录 8
Hadoop 集群节点动态管理	附录 9
Kafka 安装部署和配置	附录 10
Spark 安装部署和配置	附录 11
Flink 安装部署和配置	附录 14

4.4 任务三：使用 Spark 抽取离线数据

使用 Spark 抽取离线数据的流程如图 4.3 所示。

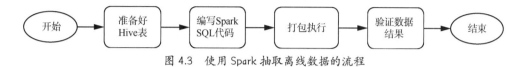

图 4.3 使用 Spark 抽取离线数据的流程

按照操作步骤，可以将 Spark 抽取离线数据的过程划分为 4 个阶段，分别是准备好 Hive 表、编写 Spark SQL 代码、打包执行和验证数据结果。

我们要抽取的原始数据存储在 MySQL 中，数据库为 test，表为 info，结构为 info(id string,city string,mode string,time string,deposit float,type string)。其中 id 为业务编号，如 000001；city 为所在城市，如北京；mode 为业务办理的方式，如柜台、自动存取款机；time 为业务办理的时间，如 2022-05-01 12:20:12；deposit 为个人存款，如 121314.50 元；type 为业务类型，如存款、取款、转账、贷款、信用卡等业务。数据如下所示。

```
000001,北京,柜台,2022-05-01 12:20:12,121314.50,转账
000002,北京,柜台,2022-05-02 13:02:31,2222.00,存款
000003,上海,自动存取款机,2022-05-03 09:22:15,3333.00,取款
```

读者可以自行录入一定的测试数据。

值得注意的是，实际金融业务数据比较多且复杂，有可能是数量庞大、相互关联的多个数据表，可以事先将需要的数据提取到一张表（如本项目），也可以在 Spark 抽取时用复杂的查询或多条 SQL 来完成。

下面我们使用 Spark SQL 来抽取 MySQL 中的离线数据，并写入 Hive 表中。

1．准备好 Hive 数据库和表

```
shiyanlou:~/ hive
hive> create database t1;
hive> use t1;
hive> create table info(id string,city string,mode string,time string,deposit float,type string) row format delimited fields terminated by ",";
```

2．编写 Spark SQL 代码

首先添加 Spark SQL 的依赖。

```xml
<dependency>
    <groupId>org.apache.spark</groupId>
    <artifactId>spark-sql_2.11</artifactId>
    <version>2.1.1</version>
</dependency>
```

然后编写 Spark SQL 完成数据的抽取,其完整代码如程序清单 4.1 所示。

程序清单 4.1

```scala
package org.lanqiao.bigdata

import org.apache.spark.sql.SparkSession
import org.apache.spark.sql.Row
import java.util.Properties
import org.apache.spark.sql.SaveMode

object SparkJdbc {
  def main(args:Array[String]):Unit={
    //Spark 2.0 开始使用 spark 来代替之前的 SparkContext 和 SQLContext
val spark=SparkSession.builder.appName("SparkJdbc").getOrCreate
    // 配置分区数和内存
spark.conf.set("spark.sql.shuffle.partitions",2)
spark.conf.set("spark.executor.memory","1g")
    // 设置 MySQL 登录名和密码
val props=new Properties();
props.put("user","root");
props.put("password","123456");
    // 连接 JDBC, 查询 MySQL 中的 info 表, 生成 DataFrame
valdf=spark.read.jdbc("jdbc:mysql://192.168.128.131:3306/test","info",props)
    // 注册临时表 t_info, 与数据集绑定 (来自 MySQL)
df.registerTempTable("t_info")
    // 从临时表查询所有数据, 插入 Hive 表 info 中
spark.sql("use t1")
spark.sql("insert into info select * from t_info")
// 关闭 spark
spark.stop()
  }
}
```

3．打包执行

在提交任务之前，因为需要连接 JDBC，所以需要把 MySQL 的连接驱动放到 Spark 的 jars 目录中，确保集群中每个节点 spark-2.1.1/jars/ 下都连接了驱动 jar 包。

```
[hadoop@master ~]$ cp /soft/hive-2.3.4/lib/mysql-connector-java-5.1.35-bin.jar /soft/spark-2.1.1/jars/
```

将程序打成 jar 包 SparkJdbc.jar，上传到 Spark 集群，提交 Spark 任务。

```
[hadoop@master ~]$ spark-submit --master yarn --deploy-mode cluster \
--class org.lanqiao.bigdata.SparkJdbc --executor-memory 1g \
/home/hadoop/SparkJdbc.jar
```

4．验证数据结果

Spark 程序执行后，访问 Hive 中的表 t1.info，结果如下。

```
hive> use t1;
hive> select * from info limit 10;
000001,北京,柜台,2022-05-01 12:20:12,121314.50,转账
000002,北京,柜台,2022-05-02 13:02:31,2222.00,存款
000003,上海,自动存取款机,2022-05-03 09:22:15,3333.00,取款
000004,天津,自动存取款机,2022-05-04 14:54:51,4444.00,取款
000005,成都,柜台,2022-05-05 15:03:56,5555.00,贷款
000006,上海,柜台,2022-05-06 10:31:23,6666.00,信用卡
......
```

由结果可知，内容符合要求。

4.5 任务四：使用 Spark 统计离线数据

使用 Spark 统计离线数据的流程如图 4.4 所示。

图 4.4 使用 Spark 统计离线数据的流程

按照操作步骤，可以将 Spark 统计离线数据的过程划分为 5 个阶段，分别是 Spark 数据清洗、Spark 数据分析、Spark 数据存储到 MySQL、打包执行和验证结果。其中前 3 步合并到一起。

1. Spark 清洗、分析和存储数据

首先添加 Spark SQL 的依赖。

```
<dependency>
    <groupId>org.apache.spark</groupId>
    <artifactId>spark-sql_2.11</artifactId>
    <version>2.1.1</version>
</dependency>
```

登录 MySQL，假设 MySQL 数据库账号为 hive，密码为 hive，然后在 MySQL 中数据库 test 下创建 2 张数据表 mode 和 city，每张表都含有 2 个字段：一个是 name，字符串类型；另一个是 num，整数类型。mode 表用来存储各种业务办理方式的数量，city 表用来存储各个城市的业务量。

```
[hadoop@master ~]$ mysql -u hive -p
Enter password:
```

输入密码"hive"。

```
mysql> use test;
mysql> create table mode(name varchar(100),num int);
mysql> create table city(name varchar(100),num int);
```

然后编写 Spark SQL 完成数据的离线统计，其完整代码如程序清单 4.2 所示。

程序清单 4.2

```scala
package org.lanqiao.bigdata

import org.apache.spark.sql.SparkSession
import org.apache.spark.sql.SaveMode
import java.util.Properties

object SparkHive {
  def main(args:Array[String]):Unit={
    // 启动 Hive 支持
val spark=SparkSession.builder.appName("SparkHive").enableHiveSupport().getOrCreate
    // 配置分区数和内存
spark.conf.set("spark.sql.shuffle.partitions",2)
spark.conf.set("spark.executor.memory","1g")
    // 使用 Hive 中数据库 t1
spark.sql("use t1")
    // 数据清洗：忽略 city 和 mode 字段值为空的数据
    // 数据分析：1. 统计各种业务办理方式的数量
```

```
    val df1=spark.sql("select mode,count(*) as num from info where city!='' and
mode!='' group by mode")
      // 数据分析：2. 统计各个城市的业务量
    val df2=spark.sql("select city,count(*) as num from info where city!='' and
mode!='' group by city")
    // 把数据分析结果以覆盖的方式插入 MySQL
    val props=new Properties();
    props.put("user","root");
    props.put("password","123456");

df1.write.mode(SaveMode.Overwrite).jdbc("jdbc:mysql://192.168.128.131:3306/
test","mode",props)
df2.write.mode(SaveMode.Overwrite).jdbc("jdbc:mysql://192.168.128.131:3306/
test","city",props)
      // 关闭 spark
    spark.stop()
    }
}
```

2．打包执行

在提交任务之前，需要把 hive-2.3.4/conf 下的 hive-site.xml 复制到 $SPARK-HOME/conf 目录下，确保每个节点在 $SPARK-HOME/conf 下都有配置文件。

```
[hadoop@master ~]$ cp /soft/hive-2.3.4/conf/hive-site.xml /soft/spark-2.1.1/
conf/
```

将程序打成 jar 包 SparkHive.jar，上传到 Spark 集群，提交 Spark 任务。

```
[hadoop@master ~]$ spark-submit --master yarn --deploy-mode cluster \
--class org.lanqiao.bigdata.SparkHive --executor-memory 1g \
/home/hadoop/SparkHive.jar
```

3．验证结果

Spark 程序执行后，访问 MySQL 中的表 test.mode 和 test.city，结果如下。

```
mysql> use test;
mysql> select * from mode;
柜台 12378
自动存取款机 31253
手机银行 3213
电话 413
其他 1025
```

```
mysql> select * from city;
北京 2345
上海 2321
湖南 2294
四川 2311
云南 1273
……
```

由结果可知，内容符合要求。

4.6 任务五：使用 Flume+Kafka 实时采集数据

使用 Flume+Kafka 实时采集数据的流程如图 4.5 所示。

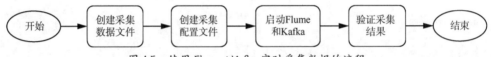

图 4.5 使用 Flume+Kafka 实时采集数据的流程

按照操作步骤，可以将 Flume+Kafka 实时采集数据的过程划分为 4 个阶段，分别是创建采集数据文件、创建采集配置文件、启动 Flume 和 Kafka 以及验证采集结果。

1．创建采集数据文件

在 ~ 目录下新建文件 data.log，该文件用来模拟产生银行业务数据（我们会向文件中写入业务数据）。

```
[hadoop@master~]$ touch data.log
```

2．创建采集配置文件

在 ~ 目录下新建 Flume 的配置信息文件 exec.conf，完整代码如程序清单 4.3 所示。

程序清单 4.3
```
[hadoop@master~]$ vi exec.conf
a1.sources=r1
a1.channels=c1
a1.sinks=k1

a1.sources.r1.type=exec
a1.sources.r1.command=tail -F /home/hadoop/data.log
a1.sources.r1.channels=c1
```

```
a1.channels.c1.type=memory
a1.channels.c1.capacity=10000
a1.channels.c1.transactionCapacity=100

a1.sinks.k1.type=org.apache.flume.sink.kafka.KafkaSink
# 设置 Kafka 的主题 topic
a1.sinks.k1.topic=bank
# 设置消费者编码为 UTF-8
a1.sinks.k1.custom.encoding=UTF-8
# 绑定 Kafka 主机以及端口号
a1.sinks.k1.brokerList=192.168.128.131:9092,192.168.128.132:9092,192.168.128.133:9092
# 设置 Kafka 序列化方式
a1.sinks.k1.serializer.class=kafka.serializer.StringEncoder
a1.sinks.k1.requiredAcks=1
a1.sinks.k1.batchSize=20
a1.sinks.k1.channel=c1
```

3. 启动 Flume 和 Kafka 服务

启动 Flume 服务。

```
[hadoop@master soft]$ flume-ng agent -n a1 -f /home/hadoop/exec.conf -Dflume.root.logger=INFO,console
```

启动 Kafka 服务。

```
[hadoop@master soft]$ kafka-server-start.sh -daemon $KAFKA_HOME/config/server.properties
```

4. 验证采集结果

在 Kafka 中创建一个名为 bank 的主题。

```
[hadoop@master soft]$ kafka-topics.sh --create --zookeeper 192.168.128.131:2181 --replication-factor 2 --partitions 3 --topic bank
```

启动 Kafka 客户端消费者,准备消费主题 bank 的消息。

```
[hadoop@slave1 soft]$ kafka-console-consumer.sh --bootstrap-server 192.168.128.131:9092 --from-beginning --topic bank --group ppp
```

使用重定向往 /home/hadoop/data.log 文件中追加内容。

```
[hadoop@master ~]$ echo 'hello'>> data.log
```

可以看到该条消息被 Kafka 消费者消费了。

至此，Flume 和 Kafka 的配置已经完成。

4.7　任务六：使用 Flink 实时计算数据

使用 Flink 实时计算数据的流程如图 4.6 所示。

图 4.6　使用 Flink 实时计算数据的流程

按照操作步骤，可以将 Flink 实时计算数据的过程划分为 4 个阶段，分别是分析数据、编写 Flink 程序、打包执行和验证数据结果。

首先我们来看原始的数据格式。

{"id":"000001","city":"北京","mode":"柜台","time":"2022-05-01 12:20:12","deposit":121314.50,"type":"转账"}

每一条数据都代表一个业务，id 表示业务编号，字符串类型，如 "000001"；city 表示业务所在城市，字符串类型，如 " 北京 "；mode 表示业务办理方式，字符串类型，如 " 柜台 "；time 表示业务发生时间，字符串类型，如 "2022-05-01 12:20:12"；deposit 表示用户存款，浮点数类型，如 121314.50；type 表示业务类型，字符串类型，如 " 转账 "。

针对需求：统计分析银行截止到目前，每个月的业务量。

（1）分析数据。

该需求需要按照月份来分段汇总数量，在 Redis 中使用字符串类型的变量 business01、business02……business12 来分别代表 12 个月各自的业务量。假设当前只有前 10 个月的数据，那么前 10 个变量值不为 0，后面 2 个变量值为 0。同时在 MySQL 数据库中创建一张表 business(month varchar(10),num int)，该表用来存储每个月的业务量，其中 month 用来存储月份，字符串类型；num 用来存储业务量，整数类型。

```
mysql> create table business(month varchar(10),num int);
```

（2）编写 Flink 程序。

首先引入相关依赖。

```
<!-- 解析JSON的包 -->
<dependency>
    <groupId>com.alibaba</groupId>
```

```xml
        <artifactId>fastjson</artifactId>
        <version>1.2.73</version>
</dependency>
<!-- 解析 Redis 的包 -->
<dependency>
        <groupId>redis.clients</groupId>
        <artifactId>jedis</artifactId>
        <version>3.2.0</version>
</dependency>
<!-- 解析 Flink 的包 -->
<dependency>
        <groupId>org.apache.flink</groupId>
        <artifactId>flink-scala</artifactId>
        <version>1.10.2</version>
</dependency>
<dependency>
        <groupId>org.apache.flink</groupId>
        <artifactId>flink-streaming-scala_2.11</artifactId>
        <version>1.10.2</version>
</dependency>
<dependency>
        <groupId>org.apache.flink</groupId>
        <artifactId>flink-connector-kafka-0.11_2.11</artifactId>
        <version>1.10.2</version>
</dependency>
```

然后编写 Flink 代码实时计算，它的主要逻辑是实时地获取 Kafka 中的数据，然后根据业务需求，计算每个月的业务量，再分别保存到 Redis 和 MySQL 中，完整代码如程序清单 4.4 所示。

程序清单 4.4

```
package org.lanqiao.bigdata

import com.alibaba.fastjson.JSON
import java.sql.{Connection,DriverManager,PreparedStatement}
import org.apache.flink.api.common.restartstrategy.RestartStrategies
import org.apache.flink.streaming.api.CheckpointingMode
import org.apache.flink.streaming.api.environment.CheckpointConfig
import org.apache.flink.streaming.api.scala.DataStream
import org.apache.kafka.clients.consumer.ConsumerConfig
import org.apache.flink.configuration.Configuration
```

```scala
import org.apache.flink.streaming.api.datastream.DataStreamSink
import org.apache.flink.streaming.api.functions.sink.RichSinkFunction
import org.apache.flink.streaming.api.scala.StreamExecutionEnvironment
import java.util.Properties
import org.apache.flink.streaming.util.serialization.SimpleStringSchema
import org.apache.flink.streaming.connectors.kafka.FlinkKafkaConsumer
import org.apache.flink.shaded.jackson2.com.fasterxml.jackson.databind.node.ObjectNode
import org.apache.flink.streaming.api.scala._
import org.apache.flink.streaming.connectors.redis.common.config.FlinkJedisPoolConfig;

object FlinkKafka {
  def main(args:Array[String]):Unit={
    // 创建 flink 流计算执行环境
    val environment:StreamExecutionEnvironment=StreamExecutionEnvironment.getExecutionEnvironment
    // 开启 CheckPointing 可以开启重启策略
    environment.enableCheckpointing(5000)
    // 设置重启策略为，出现 3 次异常重启 3 次，隔 10 秒一次
    environment.getConfig.setRestartStrategy(RestartStrategies.fixedDelayRestart(3,10000))
    // 系统异常退出或者人为退出，不删除 checkpoint 数据
    environment.getCheckpointConfig.enableExternalizedCheckpoints(CheckpointConfig.ExternalizedCheckpointCleanup.RETAIN_ON_CANCELLATION)
    // 设置 Checkpoint 模式（与 Kafka 整合，要设置 Checkpoint 模式为 Exactly_Once）端到端的一致性
    environment.getCheckpointConfig.setCheckpointingMode(CheckpointingMode.EXACTLY_ONCE)
    // 测试环境设置并行度为 1，生产环境可以调大，也可以不设置
    environment.setParallelism(1)
    // 创建事件创建时间
    //environment.setStreamTimeCharacteristic(TimeCharacteristic.EventTime)
    // 配置 Redis
    val conf=new FlinkJedisPoolConfig.Builder().setHost("192.168.128.131").setPort(6379).build()
    // 配置 Kafka 信息
    val properties=new Properties()
    // 配置 Kafka IP 地址以及 Kafka 端口号
    properties.setProperty(ConsumerConfig.BOOTSTRAP_SERVERS_CONFIG,"192.168.128.131:9092,192.168.128.132:9092,192.168.128.133:9092")
```

```
    // 设置 Kakfa 消费者组名称
properties.setProperty(ConsumerConfig.GROUP_ID_CONFIG,"mygroup")
    // 开启序列化配置
properties.setProperty(ConsumerConfig.KEY_DESERIALIZER_CLASS_CONFIG,"org.apache.kafka.common.serialization.StringSerializer")
    // 开启反序列化配置
properties.setProperty(ConsumerConfig.VALUE_DESERIALIZER_CLASS_CONFIG,"org.apache.kafka.common.serialization.StringDeserializer")
    // 从最新消息进行消费: latest
properties.setProperty(ConsumerConfig.AUTO_OFFSET_RESET_CONFIG,"latest")
    //Kafka 的消费者，不自动提交偏移量
properties.setProperty(ConsumerConfig.ENABLE_AUTO_COMMIT_CONFIG,"false")
    // 添加数据来源为 Kafka
val topic:DataStream[ObjectNode]=environment.addSource(new FlinkKafkaConsumer[ObjectNode]("bank",new SimpleStringSchema(),properties))
    /**
     * 打印输出 Kafka 接收到的 JSON 数据
     *{"id":"000001","city":"北京","mode":"柜台",
     *"time":"2022-05-01 12:20:12","deposit":121314.50,"type":"转账"}
     */
    //topic.map(t=>t.toString).print()
valdataStream:DataStreamSink[(String,Int)]=
topic.map(t=>{
val month=JSON.parseObject(t).getString("time").substring(5,7)
        (month,1)
    })
    // 根据 month 进行分组
    .keyBy(0)
    // 求和
    .sum(1)
    // 打印
    .print()

    // 保存数据到 Redis
dataStream.addSink(new RedisSink[(string,int)](conf,new MyRedisMapper))
    // 保存数据到 MySQL
dataStream.addSink(new MySQLSink)

    // 提交 Flink 任务 job
environment.execute()
    }
}
```

```scala
class MyRedisMapper extends RedisMapper[(String,Int)](){
  //设置写入Redis方式
  override def getCommandDescription:RedisCommandDescription=
    new RedisCommandDescription(RedisCommand.SET)
  //key 的字段
  override def getKeyFromData(t:(String,Int)):String="business"+t._1
  //value 的字段
  override def getValueFromData(t:(String,Int)):Int=t._2
}

class MySQLSink extends RichSinkFunction[(String,Int)]{
  private var connection:Connection=null
  private var ps:PreparedStatement=null

  override def open(parameters:Configuration):Unit={
    Class.forName("com.mysql.jdbc.Driver")
    connection=DriverManager.getConnection("jdbc:mysql://192.168.128.131:3306/test ","root","123456")
  }

  override def invoke(value:(String,Int)):Unit={
    try {
      ps=connection.prepareStatement("select * from business where month=?")
      ps.setString(1,value._1)
      ResultSet rs=ps.executeQuery();
      if(rs.next()) {//存在则更新
        ps=connection.prepareStatement("update business set num=? where month=?")
        ps.setInt(1,value._2)
        ps.setString(2,value._1)
        ps.executeUpdate()
      }else{//不存在则插入
        ps=connection.prepareStatement("insert into business values(?,?)")
        ps.setString(1,value._1)
        ps.setInt(2,value._2)
        ps.executeUpdate()
      }
    } catch {
      case e:Exception=>println(e.getMessage)
    }
  }
}
```

```
    // 关闭连接操作
    override def close():Unit={
      if (connection!=null) {
          connection.close()
      }
      if (ps!=null){
          ps.close()
      }
    }
}
```

（3）打包执行。

将程序打成 jar 包 FlinkKafka.jar，上传到 Flink 集群执行。

```
[hadoop@master ~]$/soft/flink-1.10.2/bin/flink run -c org.lanqiao.bigdata.FlinkKafka
FlinkKafka.jar
```

启动后，我们编写的程序就能获取 bank 主题下的消息了。

（4）验证数据结果。

当我们使用重定向往 /home/hadoop/data.log 文件中追加如下银行业务数据时：

```
[hadoop@master ~]$ echo "{'id':'000001','city':'北京','mode':'柜台','time':
'2022-01-01 12:20:12','deposit':3213.50,'type':'存款'}">> data.log
    [hadoop@master ~]$ echo "{'id':'000002','city':'上海','mode':'柜台','time':
'2022-01-02 12:20:12','deposit':12361.00,'type':'取款'}">> data.log
    [hadoop@master ~]$ echo "{'id':'000003','city':'天津','mode':'自动存取款机',
'time':'2022-02-03 12:20:12','deposit':6789.00,'type':'转账'}">> data.log
    [hadoop@master ~]$ echo "{'id':'000004','city':'北京','mode':'柜台','time':
'2022-02-04 12:20:12','deposit':12839.00,'type':'取款'}">> data.log
    [hadoop@master ~]$ echo "{'id':'000005','city':'成都','mode':'自动存取款机',
'time':'2022-01-03 12:20:12','deposit':314.50,'type':'存款'}">> data.log
```

这些业务数据会被采集到 Kafka 中，然后 Flink 开始消费这些 JSON 格式的消息并计算。将结果打印出来，我们会看到如下输出。

```
……
(01,1)
(01,2)
(02,1)
(02,2)
(01,3)
……
```

数据被实时地计算出来，同时写入 Redis 和 MySQL 数据库中。

4.8 任务七：Vue.js+Java+ECharts 数据可视化

Vue.js+Java+ECharts 数据可视化的流程如图 4.7 所示。

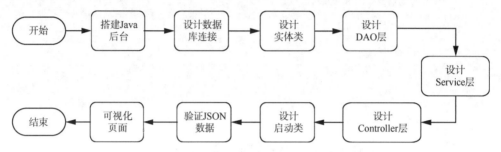

图 4.7　Vue.js+Java+ECharts 数据可视化的流程

按照操作步骤，可以将 Vue.js+Java+ECharts 数据可视化的过程划分为 9 个阶段，分别是搭建 Java 后台、设计数据库连接、设计实体类、设计 DAO 层、设计 Service 层、设计 Controller 层、设计启动类、验证 JSON 数据和可视化页面。

针对需求一：统计分析银行各种业务办理方式的占比，并采用 ECharts 饼图可视化。

（1）搭建 Java Web 后台开发框架（SpringBoot）。

创建 SpringBoot 项目，操作过程如下。

① 打开 Eclipse，新建 Spring Boot → Spring Starter Project，单击"next"。

② 选择 jar 方式，设置 java 为"java8"，group 为"org.lanqiao"，Artifact 为"bigdata"，Package 为"org.lanqiao.bigdata"，单击"next"。

③ 选择事先要加载的组件。分别选择 SQL 里的"MyBatis Framework"和"MySQL Driver"，NoSQL 里的"Spring Data Redis（Access+Driver）"，Web 里的"Spring Web"，依次单击"next"和"finish"。

④ 单击操作完成后，需要稍等一会儿，maven 会自动下载相关 jar 包资源。

（2）设计数据库连接。

打开 src/main/resources 目录下的 application.properties 文件，代码如程序清单 4.5 所示。

程序清单 4.5

```
spring.datasource.url=jdbc:mysql://localhost:3306/test
spring.datasource.username=hive
spring.datasource.password=hive
spring.datasource.driver-class-name=com.mysql.cj.jdbc.Driver
```

```
# REDIS (RedisProperties)
# Redis 数据库索引（默认为 0）
spring.redis.database=0
# Redis 服务器地址
spring.redis.host=localhost
# Redis 服务器连接端口
spring.redis.port=6379
# Redis 服务器连接密码（默认为空）
spring.redis.password=
# 连接池最大连接数（使用负值表示没有限制）
spring.redis.pool.max-active=8
# 连接池最大阻塞等待时间（使用负值表示没有限制）
spring.redis.pool.max-wait=-1
# 连接池中的最大空闲连接
spring.redis.pool.max-idle=8
# 连接池中的最小空闲连接
spring.redis.pool.min-idle=0
# 连接超时时间（毫秒）
spring.redis.timeout=0
```

（3）设计实体类。

创建包 org.lanqiao.bigdata.bean，然后在该包下创建类 Mode，代码如程序清单 4.6 所示。

程序清单 4.6

```java
package org.lanqiao.bigdata.bean;

public class Mode{
    private String name;
    private int num;

    public Mode(String name,int num){
        super();
        this.name=name;
        this.num=num;
    }

    public String getName(){
        return name;
    }
    public void setName(String name){
        this.name=name;
```

```java
    }
    public int getNum(){
        return num;
    }
    public void setNum(int num){
        this.num=num;
    }

    @Override
    public String toString(){
        return "Mode [name=" + name + ",num=" + num + "]";
    }
}
```

（4）设计 DAO 层。

创建包 org.lanqiao.bigdata.dao，然后在该包下创建接口 ModeMapper，代码如程序清单 4.7 所示。

程序清单 4.7

```java
package org.lanqiao.bigdata.dao;

import java.util.List;
import org.apache.ibatis.annotations.Mapper;
import org.apache.ibatis.annotations.Select;
import org.lanqiao.bigdata.bean.Mode;
import org.springframework.stereotype.Repository;

@Mapper
@Repository
public interface ModeMapper{
    @Select("SELECT name,num FROM mode order by num desc")
    List<Mode>getMode();
}
```

（5）设计 Service 层。

创建包 org.lanqiao.bigdata.service，然后在该包下创建接口 ModeService，代码如程序清单 4.8 所示。

程序清单 4.8

```java
package org.lanqiao.bigdata.service;
```

```
import java.util.List;
import org.lanqiao.bigdata.bean.Mode;

public interface ModeService {
    List<Mode>getMode();
}
```

再创建实现类 ModeServiceImpl,代码如程序清单 4.9 所示。

程序清单 4.9

```
package org.lanqiao.bigdata.service;

import java.util.List;
import org.lanqiao.bigdata.bean.Mode;
import org.lanqiao.bigdata.dao.ModeMapper;
import org.springframework.beans.factory.annotation.Autowired;
import org.springframework.stereotype.Service;

@Service
public class ModeServiceImpl implements ModeService {
    @Autowired
    ModeMappermodeMapper;

    @Override
    public List<Mode>getMode() {
        return modeMapper.getMode();
    }
}
```

(6)设计 Controller 层。

创建包 org.lanqiao.bigdata.controller,然后在该包下创建类 ModeController,代码如程序清单 4.10 所示。

程序清单 4.10

```
package org.lanqiao.bigdata.controller;

import java.util.ArrayList;
import java.util.List;
import java.util.Map;
import java.util.HashMap;
import org.lanqiao.bigdata.bean.Mode;
import org.lanqiao.bigdata.service.ModeService;
```

```java
import org.springframework.beans.factory.annotation.Autowired;
import org.springframework.web.bind.annotation.RequestMapping;
import org.springframework.web.bind.annotation.RestController;

@RestController
public class ModeController {
    @Autowired
    ModeService modeService;

    @RequestMapping("/mode")
    public List<Map<String,Object>>getMode(){
        List<Mode> list=modeService.getMode();
        List<Map<String,Object>> li=new ArrayList<Map<String,Object>>();
        for(int i=0;i<list.size();i++) {
            Mode m=list.get(i);
            Map<String,Object> map=new HashMap<String,Object>();
            map.put("name",m.getName());
            map.put("value",m.getNum());
            li.add(map);
        }
        return li;
    }
}
```

(7) 设计启动类。

在包 org.lanqiao.bigdata 下创建启动类 DemoApplication, 代码如程序清单 4.11 所示。

程序清单 4.11

```java
package org.lanqiao.bigdata;

import org.springframework.boot.SpringApplication;
import org.springframework.boot.autoconfigure.SpringBootApplication;
import org.springframework.context.annotation.ComponentScan;

@SpringBootApplication
@ComponentScan(basePackages={"org.lanqiao.bigdata"})
public class DemoApplication {
    public static void main(String[] args) {
        SpringApplication.run(DemoApplication.class,args);
    }
}
```

(8)验证 JSON 数据。

右击项目→选择"run as"→选择"spring boot app",打开浏览器,输入地址"http://localhost:8080/mode",查看 JSON 数据是否如下所示。

```
[{"name":"自动存取款机","value":"31253"},{"name":"柜台","value":"12378"},{"name":"手机银行","value":"3213"},{"name":"其他","value":"1025"},{"name":"电话","value":"413"}]
```

(9)可视化页面。

下面按照需求用 ECharts 饼图来呈现。

在项目 static 目录下创建 mode.vue,代码如程序清单 4.12 所示。

程序清单 4.12

```
<template>
<div class="chart-container">
        <div id="chartPie" style="width:100%; height:550px;"></div>
</div>
</template>

<script>
exportdefault{
        name:"Typechart",
        data(){
        return{
            chartPie:'',
            }
        },
        methods:{
    drawPieChart(){
            //基于准备好的dom,初始化ECharts实例
            this.chartPie=this.$echarts.init(document.getElementById("chartPie"));
            this.chartPie.setOption({
                //设置标题、副标题,以及标题位置居中
                title:{
                    text:'银行各种业务办理方式占比图',
                    x:'center'
                },
                //具体点击某一项触发的样式内容
                tooltip:{
                    trigger:'item',
```

```
                        formatter:"{b}:{c} 次 ({d}%)"
                },
                // 左上侧分类条形符
                legend:{
                        orient:'vertical',
                        left:'left'
                },
                // 饼状图类型以及数据源
                series:[
                        {
                                name:' 统计数量 ',
                                type:'pie',
                                center:['30%','70%'],
                                // 通过跨域获取数据给 data 赋值
                                data:[],
                                // 设置饼状图扇形区域的样式
                                itemStyle:{
                                        emphasis:{
                                                shadowBlur:10,
                                                shadowOffsetX:0,
                                                shadowColor:'rgba(0,0,0,0.5)'
                                        }
                                },
                        }
                ]
        });
},
// 动态获取饼状图的数据
async initData(){
        this.axios.post("mode",{}).then(res=>{
                var getData=[];
                //for 循环赋值
                for(let i=0;i<res.data.length;i++){
                        var obj=newObject();
                        obj.name=res.data[i].name;
                        obj.value=res.data[i].value;
                        getData[i]=obj;
                }
                // 然后给饼状图赋值
                this.chartPie.setOption({
                        series:[{
                                data:getData,
                        }]
```

```
                    });
                })
            },
            drawCharts(){
                this.drawPieChart();
            },
        },
        // 页面一加载就调用方法
        mounted (){
        // 先调用这个方法赋值
            this.initData();
        // 再调用饼状图方法
            this.drawCharts();
        }
    }
</script>

<style scoped>
    .chart-container {
        width:100%;
        float:left;
    }
</style>
```

代码编写完毕，项目启动并运行后，打开浏览器，输入地址"http://localhost:8080/mode.html"，就能看到可视化数据了，如图 4.8 所示。

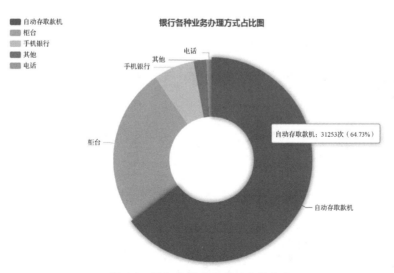

图 4.8　银行各种业务办理方式占比

针对需求二：统计分析银行各个城市的业务量，并采用 ECharts 柱状图可视化。

（1）搭建 Java Web 后台开发框架（SpringBoot）。

项目后台之前已经搭建完毕。

（2）设计数据库连接。

数据库连接参数保持不变。

（3）设计实体类。

创建包 org.lanqiao.bigdata.bean，然后在该包下创建类 City，代码如程序清单 4.13 所示。

程序清单 4.13

```java
package org.lanqiao.bigdata.bean;

public class City{
    private String name;
    private int num;

    public City(String name,int num){
        super();
        this.name=name;
        this.num=num;
    }

    public String getName(){
        return name;
    }
    public void setName(String name){
        this.name=name;
    }
    public int getNum(){
        return num;
    }
    public void setNum(int num){
        this.num=num;
    }

    @Override
    public String toString(){
        return "City [name="+name+",num="+num+"]";
    }
}
```

（4）设计 DAO 层。

创建包 org.lanqiao.bigdata.dao，然后在该包下创建接口 CityMapper，代码如程序清单 4.14 所示。

程序清单 4.14
```java
package org.lanqiao.bigdata.dao;

import java.util.List;
import org.apache.ibatis.annotations.Mapper;
import org.apache.ibatis.annotations.Select;
import org.lanqiao.bigdata.bean.City;
import org.springframework.stereotype.Repository;

@Mapper
@Repository
public interface CityMapper {
    @Select("SELECT name,num FROM city order by num desc")
    List<City> getCity();
}
```

（5）设计 Service 层。

创建包 org.lanqiao.bigdata.service，然后在该包下创建接口 CityService，代码如程序清单 4.15 所示。

程序清单 4.15
```java
package org.lanqiao.bigdata.service;

import java.util.List;
import org.lanqiao.bigdata.bean.City;

public interface CityService {
    List<City> getCity();
}
```

再创建实现类 CityServiceImpl，代码如程序清单 4.16 所示。

程序清单 4.16
```java
package org.lanqiao.bigdata.service;

import java.util.List;
import org.lanqiao.bigdata.bean.City;
```

```
import org.lanqiao.bigdata.dao.CityMapper;
import org.springframework.beans.factory.annotation.Autowired;
import org.springframework.stereotype.Service;

@Service
public class CityServiceImpl implements CityService {
    @Autowired
    CityMapper cityMapper;

    @Override
    public List<City> getCity(){
        return cityMapper.getCity();
    }
}
```

（6）设计 Controller 层。

创建包 org.lanqiao.bigdata.controller，然后在该包下创建类 CityController，代码如程序清单 4.17 所示。

程序清单 4.17

```
package org.lanqiao.bigdata.controller;

import java.util.ArrayList;
import java.util.List;
import java.util.Map;
import java.util.HashMap;
import org.lanqiao.bigdata.bean.City;
import org.lanqiao.bigdata.service.CityService;
import org.springframework.beans.factory.annotation.Autowired;
import org.springframework.web.bind.annotation.RequestMapping;
import org.springframework.web.bind.annotation.RestController;

@RestController
public class CityController {
    @Autowired
    CityService cityService;

    @RequestMapping("/city")
    public Map<String,Object> getCity(){
        List<City> list=cityService.getCity();
        String[] x=new String[list.size()];
```

```
            Integer[] y=new Integer[list.size()];
            for(int i=0;i<x.length;i++) {
                    City c=list.get(i);
                    x[i]=c.getName();
                    y[i]=c.getNum();
            }
            Map map=new HashMap<String,Object>();
            map.put("x",x);
            map.put("y",y);
            return map;
    }
}
```

（7）设计启动类。

启动类保持不变。

（8）验证 JSON 数据。

右击项目 → 选择 "run as" → 选择 "spring boot app"，打开浏览器，输入地址 "http://localhost:8080/city"，查看 JSON 数据是否如下所示。

{"x":["北京","上海","浙江","四川","河北","湖南","山东","江苏","广东","河南","安徽","湖北","福建","陕西","贵州","江西","重庆","山西","云南","广西","甘肃","天津","青海","辽宁","内蒙古","黑龙江","宁夏","吉林","新疆","海南","澳门","西藏"],"y":[2345,2321,2315,2311,2310,2294,2239,2217,2214,2020,1987,1965,1879,1876,1765,1678,1587,1273,1245,1143,1123,1123,876,871,654,623,542,471,457,413,387,362]}

（9）可视化页面。

下面按照需求用 ECharts 柱状图来呈现。

在项目 static 目录下创建 city.vue，代码如程序清单 4.18 所示。

程序清单 4.18

```
<template>
<div class="chart-container">
        <div id="chartBar" style="width:100%; height:550px;"></div>
</div>
</template>

<script>
exportdefault{
        name:"Typechart",
        data(){
        return{
```

```
                    chartBar:'',
                }
            },
            methods:{
                drawBarChart(){
                    // 基于准备好的 dom, 初始化 ECharts 实例
                    this.chartBar=this.$echarts.init(document.getElementById("chartBar"));
                    this.chartBar.setOption({
                        // 设置标题、副标题, 以及标题位置居中
                        title:{
                            text:' 银行各个城市的业务量 ',
                        },
                        // 具体点击某一项触发的样式内容
                        tooltip:{
                            trigger:'axis',
                            axisPointer:{
                                type:'shadow'
                            }
                        },
                        // 分类条形符
                        legend:{
                            data:[' 数量 ']
                        },
                        xAxis:{
                            type:'category',
                            axisLabel:{// 文字显示
                                interval:0,// 间隔 0 表示每一个都强制显示
                                rotate:40// 旋转角度 [-90,90]
                            },
                            data:[] // 数据由 jquery 请求后端服务器获取
                        },
                        // 设置绘图网格, 单个 grid 内最多可以放置上下两个 x 轴, 左右两个 y 轴, 设置绘图网格与左边和下边的距离
                        // 这里主要用于 axisLabel 文字显示
                        grid:{
                            left:'10%',
                            bottom:'35%'
                        },
                        //y 轴内容
                        yAxis:{
```

```
                    type:'value',
                    axisLabel:{
                        formatter:'{value}'
                    }
                },
                // 绘图类型和数据
                series:[{
                    name:' 数量 ',
                    data:[],// 数据由 jquery 请求后端服务器获取
                    type:'bar',// 柱状图
                    itemStyle:{
                        normal:{
                            color:function(params){
                                var colorList=['red','orange',
'yellow','green','blue','purple','black','#123456','#789123','gray'];
                                return colorList[params.dataIndex]
                            }
                        }
                    }
                }]
            });
        },
        // 动态获取柱状图的数据
        async initData(){
            this.axios.post("city",{}).then(res=>{
                // 然后给柱状图赋值
                this.chartBar.setOption({
                    xAxis:{
                        type:'category',
                        axisLabel:{// 文字显示
                            interval:0,// 间隔 0 表示每一个都强
                                       制显示
                            rotate:40// 旋转角度 [-90,90]
                        },
                        data:res.data.x  //x 轴数据
                    },
                    series:[ {
                        name:' 数量 ',
                        data:res.data.y,//y 轴数据
```

```
                                    type:'bar',//柱状图
                                    itemStyle:{
                                        normal:{
                                            color:function(params){
                                                var colorList=['red',
'orange','yellow','green','blue','purple','black','#123456','#789123','gray'];
                                                return colorList
[params.dataIndex]
                                            }
                                        }
                                    }
                                }]
                            });
                        })
                    },
                    drawCharts(){
                        this.drawBarChart();
                    },
                },
                // 页面一加载就调用方法
                mounted(){
                    // 先调用这个方法赋值
                    this.initData();
                    // 再调用柱状图方法
                    this.drawCharts();
                }
            }
        </script>

        <style scoped>
            .chart-container{
                width:100%;
                float:left;
            }
        </style>
```

代码编写完毕，项目启动并运行后，打开浏览器，输入地址"http://localhost:8080/city.html"，就能看到可视化数据了，如图 4.9 所示。

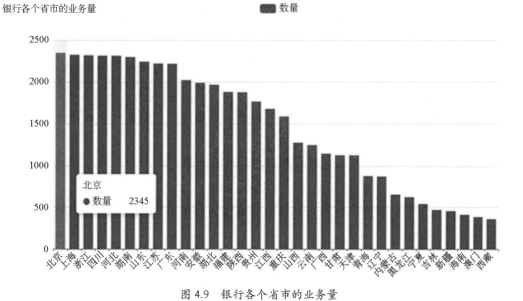

图 4.9　银行各个省市的业务量

针对需求三：实时统计分析银行截止到目前，每个月的业务量，并采用 ECharts 折线图可视化。

（1）搭建 Java Web 后台开发框架（SpringBoot）。

项目后台之前已经搭建完毕。

（2）设计数据库连接。

数据库连接参数保持不变。

（3）设计 Controller 层。

创建包 org.lanqiao.bigdata.controller，然后在该包下创建类 BusinessController，代码如程序清单 4.19 所示。

程序清单 4.19

```
package org.lanqiao.bigdata.controller;

import java.util.HashMap;
import java.util.Map;
import org.springframework.beans.factory.annotation.Autowired;
import org.springframework.data.redis.core.StringRedisTemplate;
import org.springframework.web.bind.annotation.RequestMapping;
import org.springframework.web.bind.annotation.RestController;
```

```
@RestController
public class BusinessController {
@Autowired
private static StringRedisTemplate stringRedisTemplate;

    @RequestMapping("/business")
    public Map<String,Object> getBusiness(){
        String[] x={"01","02","03","04","05","06","07","08","09","10","11","12"};
        Integer[] y=new Integer[12];
        for(int i=0;i<12;i++){
            y[i]=Integer.parseInteger(stringRedisTemplate.opsForValue().get("business"+x[i]))
        }
        Map map=new HashMap<String,Object>();
        map.put("x",x);
        map.put("y",y);
        return map;
    }
}
```

从 Redis 中取出 12 个月份变量 business01、business02……business12 对应的值，然后依次封装到数组 y、数组 x 和对象 map 中，再返回给前端调用页面。

（4）设计启动类。

启动类保持不变。

（5）验证 JSON 数据。

右击项目→选择"run as"→选择"spring boot app"，打开浏览器，输入地址"http://localhost:8080/business"，查看 JSON 数据是否如下所示。

```
{"x":["01","02","03","04","05","06","07","08","09","10","11","12"],
"y":[1212,1373,1410,1009,998,891,1025,1211,1498,1510,0,0]}
```

（6）可视化页面。

下面按照需求用 ECharts 折线图来呈现。

在项目 static 目录下创建 business.vue，代码如程序清单 4.20 所示。

程序清单 4.20

```
<template>
<div class="chart-container">
        <div id="chartLine" style="width:100%; height:550px;"></div>
```

```
</div>
</template>

<script>
exportdefault{
        name:"Typechart",
        data(){
        return{
            chartLine:'',
            }
        },
        methods:{
    drawLineChart(){
     // 基于准备好的dom,初始化ECharts实例
this.chartLine=this.$echarts.init(document.getElementById("chartLine"));
                    this.chartLine.setOption({
                        // 设置标题、副标题,以及标题位置居中
                        title:{
                text:'每个月的业务量',
                subtext:'截止到目前'
                        },
                        // 具体点击某一项触发的样式内容
                        tooltip:{
                            trigger:'axis',
                axisPointer:{
                type:'shadow'
                }
                        },
                        // 分类条形符
                        legend:{
                            data:['数量']
                        },
                        xAxis:{
                            type:'category',
                            axisLabel:{// 文字显示
                                interval:0,// 间隔0表示每一个都强制显示
                                rotate:40// 旋转角度[-90,90]
                            },
                            data:[] // 数据由jquery请求后端服务器获取
                },
```

```
                        // 设置绘图网格,单个grid内最多可以放置上下两个x轴,
左右两个y轴,设置绘图网格与左边和下边的距离
                        // 这里主要用于axisLabel文字显示
                        grid:{
                            left:'10%',
                            bottom:'35%'
                        },
                        //y轴内容
                        yAxis:{
                            type:'value',
                            axisLabel:{
                                formatter:'{value}'
                            }
                        },
                        // 绘图类型和数据
                        series:[{
                            name:'数量',
                            data:[],// 数据由jquery请求后端服务器获取
                            type:'line',// 折线图
                            itemStyle:{
                                normal:{
                                    color:function(params){
                                        var colorList=['red','orange',
'yellow','green','blue','purple','black','#123456','#789123','gray'];
                                        return colorList[params.
dataIndex]
                                    }
                                }
                            }
                        }]
                    });
                },
                // 动态获取折线图的数据
                async initData(){
                    this.axios.post("business",{}).then(res=>{
                        // 然后给折线图赋值
                        this.chartLine.setOption({
                            xAxis:{
                                type:'category',
                                axisLabel:{// 文字显示
```

```
                                        interval:0,// 间隔 0 表示
每一个都强制显示
                                        rotate:40// 旋转角度 [-90,90]
                                    },
                                    data:res.data.x //x轴数据
                                },
                                series:[ {
                                    name:' 数量 ',
                                    data:res.data.y,//y轴数据
                                    type:'line',// 折线图
                                    itemStyle:{
                                        normal:{
                                            color:function(params){
                                                var colorList=['red',
'orange','yellow','green','blue','purple','black','#123456','#789123','gray'];
                                                return colorList[params.dataIndex]
                                            }
                                        }
                                    }
                                } ]
                            });
                        })
                },
                drawCharts(){
                    this.drawLineChart();
                },
            },
            // 页面一加载就调用方法
            mounted (){
                // 先调用这个方法赋值
                this.initData();
                // 再调用折线图方法
                this.drawCharts();
            }
        }
    </script>

    <style scoped>
        .chart-container {
            width:100%;
```

```
            float:left;
        }
    </style>
```

代码编写完毕，项目启动并运行后，打开浏览器，输入地址"http://localhost:8080/business.html"，就能看到可视化数据了，如图 4.10 所示。

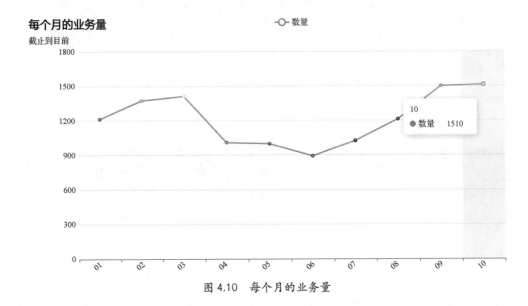

图 4.10　每个月的业务量

由于目前只有 1 月至 10 月的数据，所以后面 11 月、12 月没有数据。

4.9　任务八：使用 Spark ML 数据挖掘

我们使用的数据是历史用户购买基金的数据，其主要有 3 个属性，第一个是用户（id），string 类型；第二个是基金编号（code），string 类型；第三个是购买次数（num），int 类型。比如：

```
A10010    002190    3
A10010    002191    2
A10011    002190    1
A10011    002192    4
A10012    002190    6
……
```

以第一行为例，A10010 是客户的证券号码，代表用户；002190 是基金的编号，对应的基金名称为"农银新能源主题"；3 代表这个用户曾经购买过 3 次。

需求: 通过历史用户购买基金的数据, 使用协同过滤算法, 给当前用户推荐合适的基金。

协同过滤算法是一种较为著名和常用的推荐算法, 它基于对用户历史行为数据的挖掘发现用户的偏好, 并预测用户可能喜好的产品进行推荐。也就是常见的"猜你喜欢"和"购买了该商品的人也喜欢"等功能。

我们通过一个简单的示例, 来给大家演示其基本原理。首先我们通过合并整理用户购买基金的原始数据得到这样一个二维表。

	002191	002192	002193	002194	002195	......
A10010	1	0	2	0	3	......
A10011	1	0	0	0	0	......
A10012	0	3	0	3	3	......
A10013	1	0	1	0	1	......
A10014	0	1	1	1	0	......

第一列是用户, 第一行是基金编号, 表格中的数字代表着该用户购买该基金的次数。现在有一个新用户, 它的数据情况是:

A10015	0	3	0	2	0	......

怎么给 A10015 用户推荐基金呢? 我们发现 A10015 用户与 A10012 用户很像:

A10012	0	3	0	3	3	......

那么将 A10012 用户购买的基金推荐给 A10015 用户, A10015 用户购买的可能性较大, 基于这个假设, 我们就给 A10015 用户推荐其还没有买过的 002195 基金。

在 Spark ML 中已经集成了协同过滤算法的相关 API, 下面我们来编码实现。

首先引入相关依赖。

```xml
<dependency>
    <groupId>org.apache.spark</groupId>
    <artifactId>spark-mllib_2.11</artifactId>
    <version>2.11</version>
</dependency>
```

然后编写详细代码, 其完整代码如程序清单 4.21 所示。

程序清单 4.21

```
package org.lanqiao.bigdata

import org.apache.spark.sql.SparkSession
import org.apache.spark.ml.evaluation.RegressionEvaluator
import org.apache.spark.ml.recommendation.ALS
```

```
/**
 * 协同过滤算法
 * 输入数据为300个用户对1000个基金的购买次数
 * 数据格式:id,code,num
 * 通过用户已经购买的基金来推荐用户购买一些没有购买过的基金
 **/

// 数据格式
case class Score(id:String,code:String,num:Int)

object CollaborativeFiltering {
  def main(args:Array[String]):Unit={
val spark=SparkSession
      .builder
      .master("local[*]")
      .appName("CollaborativeFiltering")
      .getOrCreate()

    // 数据实例化
    def parseRating(str:String):Rating={
val fields=str.split(",")
      assert(fields.size==3)
Score(fields(0),fields(1),fields(2).toInt)
    }

    import spark.implicits._
    // 数据文件
val ratings=spark.read.textFile("data.txt")
      .map(parseRating)
      .toDF()
    // 80%的数据用于训练模型,20%用于测试
val Array(training,test)=ratings.randomSplit(Array(0.8,0.2))

    // 使用ALS在训练集上构建推荐模型
val als=new ALS()
      // 迭代最大值
      .setMaxIter(5)
      // ALS中正则化参数,默认为1.0
      .setRegParam(0.01)
      .setUserCol("id")
```

```
        .setItemCol("code")
        .setRatingCol("num")
    // 训练模型
    val model=als.fit(training)

    // 通过计算测试数据的均方根误差来评估模型
    // setColdStartStrategy 以不得到 NaN 值
    model.setColdStartStrategy("drop")
    val predictions=model.transform(test)

    val evaluator=new RegressionEvaluator()
        .setMetricName("rmse")
        .setLabelCol("rating")
        .setPredictionCol("prediction")
    val rmse=evaluator.evaluate(predictions)
    println(s"Root-mean-square error(均方根误差)=$rmse")

    // 为每个用户生成十大推荐基金
    val userRecs=model.recommendForAllUsers(10)
    // 为每个基金生成前 10 名用户推荐
    val codeRecs=model.recommendForAllItems(10)
    userRecs.orderBy("id").show(300,false)
    codeRecs.orderBy("code").show(1000,false)
  }
}
```

执行完毕后，部分结果如下所示。

```
id          recommendations
A10010
[[002195,4.247095],[002193,3.6386964],[002218,3.0459275],[002211,2.9969811],
[002312,2.8946023],[002381,2.4610684],[002361,2.428734],[002187,2.3305135],[002176,
2.2790494],[002198,2.2649255]]
  A10011
[[002176,2.8552358],[002298,2.5752294],[002319,2.555421],[002341,2.4367332],
[002155,2.2868218],[002376,2.2531743],[002565,2.2295506],[002563,2.160061],[002461,
2.091165],[002612,2.086698]]
  A10012
[[002567,4.980396],[002587,4.9696956],[002393,4.961116],[002339,4.917714],[002437,
4.855087],[002289,3.990784],[002434,3.9526663],[002419,3.9220076],[002440,3.8683553],
[002316,3.636133]]
```

```
A10013
    [[002451,5.0957375],[002318,4.156398],[002276,3.9523013],[002380,3.898552],
[002425,3.3176482],[002575,3.1826537],[002329,3.1740842],[002379,3.0311487],[002348,
2.9666855],[002417,2.9110801]]
A10014
    [[002353,5.5128264],[002277,4.1807184],[002352,4.0699162],[002441,4.0324464],
[002429,3.736351],[002470,3.7025547],[002362,3.6998646],[002338,3.3742704],[002287,
3.189577],[002340,3.0725355]]
 ……
code       recommendations
002191
    [[A10087,4.6209264],[A10028,4.569703],[A10182,4.247095],[A10016,3.7311819],
[A10387,3.3176482],[A10110,2.8684502],[A10112,2.7194147],[A10207,2.5982194],[A10026,
2.3475862],[A10014,2.321721]]
002192
    [[A10016,2.851448],[A10015,2.7439156],[A10100,2.428734],[A10028,2.377817],
[A10125,2.3317814],[A10311,2.1339734],[A10212,2.0025218],[A10129,1.9792455],[A10217,
1.8496197],[A10307,1.6573409]]
002193
    [[A10022,5.577275],[A10011,5.1906443],[A10113,4.805975],[A10223,4.752778],
[A10106,3.6258726],[A10082,2.875728],[A10124,2.82525],[A10325,2.1150773],[A10147,
1.9219421],[A10221,1.8820488]]
002194
    [[A10065,5.0727634],[A10026,4.240342],[A10025,3.8337395],[A10229,3.5129216],
[A10275,3.2137554],[A10196,2.5407252],[A10291,2.433772],[A10320,2.3825085],[A10311,
2.3464804],[A10315,2.302505]]
002195
    [[A10105,5.835305],[A10176,5.0880346],[A10182,4.8710938],[A10116,3.9197617],
[A10287,3.8690348],[A10142,3.736351],[A10227,3.1740842],[A10122,3.111862],[A10082,
2.704935],[A10110,2.4294906]]
```

可以看到，每个用户后面都有 10 个基金推荐，按照分值由高到低排序，分值越高越推荐。同样，每个基金后面都有 10 个潜在用户，按照分值由高到低排序，分值越高越推荐。

4.10 任务九：编写分析报告

1. 请结合数据分析结果或现实观察回答以下问题

问题1：根据数据分析结果说明银行有哪些常用的业务办理方式？哪些方式是主要的？

问题2：根据数据分析结果说明各地银行业务量的特点，为什么？

问题3：请结合现实观察简述全国银行线下网点变化趋势。

2. 问题参考答案

问题1：这里仅给出该问题答案的核心内容提示。根据需求一的分析结果，可以得出银行的业务办理方式有柜台、自动存取款机（包括自动取款机、自动存款机、自动存取款一体机）、手机银行、电话（包括人工、人工智能）以及其他方式。其中最主要的是自动存取款机（31253次）和柜台（12378次），借助自动存取款机可以进行取款、存款、转账、查询余额、修改密码等日常常用操作，使用频繁，便于银行网点附近的居民使用，是操作次数最多的设备，属于线下业务的范畴；而柜台则能进行一些复杂的操作，比如各种业务的办理——贷款、还款、代缴费、基金、保险、银证转账等，一般耗时较长。由于柜台服务窗口数量的减少，每天的业务量也在减少，在没有必要非去柜台办理的情况下，人们更多选择自助服务。

问题2：这里仅给出该问题答案的核心内容提示。各地银行业务量的特点是发达地区多（北京：2345次，上海：2321次），欠发达地区少（吉林：612次，西藏：413次），沿海多（浙江：2315次，江苏：2217次），内地少（青海：1123次，宁夏：623次），和当地经济、人口有很大关系。

问题3：这里仅给出该问题答案的核心内容提示。近年来，在利率市场化、互联网金融等大环境的变化之下，银行业开始了向智慧经营、数字化方向的转型。随着智能手机的广泛运用、数字支付工具的普及，客户对于柜台转账和现金支取的需求迅速降低。同时银行也在主动引导客户通过非接触的方式获得服务，过去银行单纯以物理网点攻城略地的模式逐渐开始式微。银行对于金融科技的投入力度不断加大，致力于通过科技手段实现降本增效，提升获客能力和品牌影响力。目前，银行已经实现了绝大部分业务的线上化，银行传统的线下网点也在大力推动通过机器自助办理业务。虽然客户对银行物理网点的依赖度有所降低，但这并不意味着银行的物理网点会被数字渠道完全取代。首先，中老年人群仍对物理网点有较强的依赖性；其次，未来的渠道发展方向是线上与线下的融合。

4.11 答疑解惑

问题1：Spark 与 Hive 有什么区别？

解答1：

（1）Spark 不是一个数据库，而是一个运行在内存中的数据分析框架；而 Hive 是一个类似 RDBMS 的数据库，并将数据存储在表结构中。

（2）Spark 通过 Spark SQL 进行数据分析，并且支持多种开发语言，如 Java、Python、Scala；而 Hive 只能通过 HiveQL 操作数据。

（3）Spark 自身无法存储数据，只能从外部数据存储中获取数据；而 Hive 基于 Hadoop 作为存储引擎，只能运行在 HDFS 之上。

（4）Spark 基于内存执行高效的数据分析以及流数据处理；而 Hive 是基于 Hadoop 的数据仓库。

问题2：Flink 与 Spark 有什么区别？

解答2：

（1）数据处理模式。

Spark 以批处理为根本，并尝试在批处理之上支持流计算。Spark Streaming 其实并不是真正意义上的"流"处理，而是"微批次"（micro-batching）处理。

Flink 则认为，流处理才是最基本的操作，批处理也可以统一为流处理。在 Flink 的世界观中，万物皆流，实时数据是标准的、没有界限的流，而离线数据则是有界限的流。

（2）数据类型。

Spark 底层数据模型是弹性分布式数据集（RDD），Spark Streaming 进行微批处理的底层接口是 DStream。

Flink 的基本数据模型是数据流（DataFlow），以及事件（Event）序列。

（3）运行架构。

Spark 做批计算，需要将任务对应的 DAG 划分为多个阶段（Stage），一个阶段完成后，经过 shuffle 再进行下一阶段的计算。而 Flink 采用的是标准的流式执行模式，一个事件在一个节点处理完后可以直接发往下一个节点进行处理。

总结：批处理可选择 Spark；实时流处理则选择 Flink。因为 Flink 的延迟是毫秒级别的，而 Spark Streaming 的延迟是秒级延迟。Flink 提供了严格的精确一次性语义保证。Flink 的窗口 API 更加灵活、语义更丰富。Flink 提供事件时间语义，可以正确处理延迟数据。Flink 还提供了更加灵活的对状态编程的 API。

问题 3：什么是 Flink 的状态机制和容错机制？

解答 3：在 Flink 中，状态和特定算子相关联。常用的就是 Managed State 下的 Keyed State。Flink 利用 checkpoint 机制对各个任务的状态进行快照保存，在故障恢复时能够保证状态一致性。Flink 通过状态后端来管理状态以及存储 checkpoint 的内容，状态后端通常支持内存、分布式文件系统、嵌入式数据库 rocksDB。

Flink 的容错机制和状态管理是息息相关的。容错本质上就是对状态的容错，就是保证状态不丢失，后续可以重新读取状态、恢复现场。

问题 4：Flink 有哪些时间类型，它们有什么区别？

解答 4：

（1）事件时间（Event Time）。事件时间是每一个独立事件在产生它的设备上发生的时间，这个时间在事件进入 Flink 以前就已经嵌入了事件，时间顺序取决于事件产生的地方，和下游数据处理系统的时间无关。

（2）接入时间（Ingestion Time）。接入时间是数据进入 Flink 系统的时间，接入时间依赖 Source Operator 所在主机的系统时钟。由于在数据接入过程生成后，接入时间的时间戳不再发生变化，和后续处理数据的 Operator 所在机器的时钟没有关系，因此不会出现某台机器时钟不一样或网络延迟致使计算结果不准确的问题。相比于 Event Time，Ingestion Time 不能处理乱序事件，所以不用生成对应的 Watermarks。

（3）处理时间（Processing Time）。处理时间是指数据在算子计算过程中获取的所在主机的系统时间。当用户选择使用 Processing Time 时，所有和时间相关的计算算子，例如 Windows 计算，在当前的任务中将直接使用其所在主机的系统时间。基于 Processing Time 时间概念，Flink 的程序性能相对较高，延迟比较低，接入系统的数据时间的相关计算彻底交给算子内部决定。虽然性能和易用性上有优点，但在处理数据乱序时，Processing Time 不是最优的选择，哪怕数据自己不乱序，若是每台机器自己的时钟不一样，也会致使数据处理过程当中出现数据乱序，Processing Time 适用于时间计算精度不是特别高的计算场景。

4.12 拓展练习

1. 拓展项目一

我们拓展一下本章项目，假设银行高层希望了解银行客户中，不同级别的存款用户有多少，各自占比是多少。假设存款在 10 万元以下的为 A 级，10 万到 100 万元的为 B 级，100 万到 1000 万元的为 C 级，1000 万到 1 亿元的为 D 级，1 亿元以上的为 E 级。于是我们给出如下需求：统计分析银行不同级别的存款用户的占比，并采用 ECharts 饼图可视化。按照本章所讲的流程来做，请同学们自行完成该任务。

2．拓展项目二

我们拓展一下本章项目，假设银行高层希望了解今年以来各个月份，该行有多少新增用户办理了个人贷款业务。于是我们给出如下需求：统计分析银行截止到目前，每个月新增的办理个人贷款业务的用户人数，并采用 ECharts 折线图可视化。按照本章所讲的流程来做，请同学们自行完成该任务。

3．全国职业院校技能大赛样卷

访问以下地址：http://www.chinaskills-jsw.org/content.jsp?id=ff8080817e9db7c1017f8fec16de0016&classid=de7bd19628f54879be3fb10f40de8767，下载附件"GZ-2022041 大数据技术与应用赛项规程 .pdf"，完成其中的样卷。

附 录

附录1 Hadoop 安装部署和配置

1. 安装 SSH

（1）查看是否已经安装 SSH。

通过 apt-cache search openssh 命令来查看是否已经安装了 SSH。

```
shiyanlou:~/ $ apt-cache search openssh
openssh-blacklist - list of default blacklisted OpenSSH RSA and DSA keys
openssh-blacklist-extra - list of non-default blacklisted OpenSSH RSA and DSA keys
openssh-client - secure shell (SSH) client,for secure access to remote machines
openssh-server - secure shell (SSH) server,for secure access from remote machines
openssh-sftp-server - secure shell (SSH) sftp server module,for SFTP access from remote machines
gesftpserver - sftp server submodule for OpenSSH
keychain - key manager for OpenSSH
libconfig-model-openssh-perl - configuration editor for OpenSsh
libghc-crypto-pubkey-openssh-dev - OpenSSH key codec
libghc-crypto-pubkey-openssh-doc - OpenSSH key codec; documentation
libghc-crypto-pubkey-openssh-prof - OpenSSH key codec; profiling libraries
libnet-openssh-compat-perl - collection of compatibility modules for Net::OpenSSH
libnet-openssh-parallel-perl - run SSH jobs in parallel
libnet-openssh-perl - Perl SSH client package implemented on top of OpenSSH
lxqt-openssh-askpass - OpenSSH user/password GUI dialog for LXQt
openssh-client-ssh1 - secure shell (SSH) client for legacy SSH1 protocol
openssh-known-hosts - download,filter and merge known_hosts for OpenSSH
razorqt-openssh-askpass - OpenSSH helper component for Razor-qt desktop environment
```

可以看到，笔者的计算机默认情况下已经安装了 SSH。

（2）安装 SSH。

如果没有安装 SSH 的话，可以直接使用 sudo apt-get install openssh-server 命令来安装

服务器端，然后使用 sudo apt-get install openssh-client 命令来安装客户端。

```
shiyanlou:~/ $ sudo apt-get install openssh-server
正在读取软件包列表... 完成
正在分析软件包的依赖关系树
正在读取状态信息... 完成
openssh-server 已经是最新版 (1:7.2p2-4ubuntu2.10)。
升级了 0 个软件包，新安装了 0 个软件包，要卸载 0 个软件包，有 0 个软件包未被升级。
shiyanlou:~/ $ sudo apt-get install openssh-client
正在读取软件包列表... 完成
正在分析软件包的依赖关系树
正在读取状态信息... 完成
openssh-client 已经是最新版 (1:7.2p2-4ubuntu2.10)。
openssh-client 已设置为手动安装。
升级了 0 个软件包，新安装了 0 个软件包，要卸载 0 个软件包，有 0 个软件包未被升级。
```

系统提示笔者的计算机中已经安装过 SSH，而且已经是最新版本了。

（3）验证是否能够登录。

SSH 安装好后，可以使用 ssh localhost 命令来模拟远程登录本机。没有设置免密之前，需要输入用户的密码来登录。设置免密之后，无须输入用户密码就可以直接登录。

2．配置本机 SSH 免密登录

（1）生成秘钥。

使用 ssh-keygen -t rsa 命令来生成秘钥，一直按回车键（按 3 次回车键）直到结束。

```
shiyanlou:~/ $ ssh-keygen -t rsa
Generating public/private rsa key pair.
Enter file in which to save the key (/home/shiyanlou/.ssh/id_rsa):
Enter passphrase (empty for no passphrase):
Enter same passphrase again:
Your identification has been saved in /home/shiyanlou/.ssh/id_rsa.
Your public key has been saved in /home/shiyanlou/.ssh/id_rsa.pub.
The key fingerprint is:
SHA256:1U6Hq/kTu59oqZh7zDH6aYVHhejYegVL/pyxmWXwMZkshiyanlou@f1019f8b2660
The key's randomart image is:
+---[RSA 2048]----+
|           . . o |
|          +.o.E  |
|         *.o++.o |
|        ..=o+o+  |
|         S. *oO  |
```

```
|        .  =o@      |
|           =o= +    |
|          .o=o=. .  |
|          +=+o++o   |
+----[SHA256]-----+
```

（2）将公钥追加到文件中。

使用 cat ~/.ssh/id_rsa.pub >> ~/.ssh/authorized_keys 命令，把生成的公钥追加到认证文件认证 authorized_keys 中。

```
shiyanlou:~/ $ cat ~/.ssh/id_rsa.pub >> ~/.ssh/authorized_keys
```

（3）给认证文件授予权限。

使用 chmod 600 ~/.ssh/authorized_keys 命令来授予权限。

```
shiyanlou:~/ $ chmod 600 ~/.ssh/authorized_keys
```

（4）验证本机是否能够免密登录。

再次使用 ssh localhost 命令来验证，发现无须输入密码就可以直接登录了。

```
shiyanlou:~/ $ ssh localhost
Welcome to Ubuntu 16.04.7 LTS (GNU/Linux 4.18.0-21-generic x86_64)
```

3．安装 JDK

Java JDK 既可以通过 apt-get 在线安装，也可以去官网下载 JDK 压缩包安装。下面我们讲解第二种安装方式。

（1）查看本机是否安装了 JDK。

使用 java -version 命令查看本机当前环境是否已经安装 JDK。

```
shiyanlou:~/ $ java -version
openjdk version "1.8.0_292"
OpenJDK Runtime Environment (build 1.8.0_292-8u292-b10-0ubuntu1~16.04.1-b10)
OpenJDK64-Bit Server VM (build 25.292-b10,mixed mode)
```

可以看到笔者的计算机中已经安装了 OpenJDK 1.8 版本。

（2）删除 OpenJDK。

如果 Ubuntu 环境中已经安装过 OpenJDK，可以使用 sudo apt-get purge openjdk* 命令将其删除。

```
shiyanlou:~/ $ sudo apt-get purge openjdk*
```

（3）下载 JDK。

前往 Oracle Java 官网下载 JDK，这里下载的版本为 jdk-8u311-linux-x64.tar.gz。也

可以使用 wget 命令下载蓝桥云课提供的资源（https://labfile.oss.aliyuncs.com/courses/7758/jdk-8u311-linux-x64.tar.gz）。

```
shiyanlou:~/ $ wget https://labfile.oss.aliyuncs.com/courses/7758/jdk-8u311-linux-x64.tar.gz
```

（4）解压缩。

创建目录 sudo mkdir /usr/lib/jvm。

```
shiyanlou:~/ $ sudo mkdir /usr/lib/jvm
```

解压缩到该目录：sudo tar -zxvf jdk-8u311-linux-x64.tar.gz -C /usr/lib/jvm。

```
shiyanlou:~/ $ sudo tar -zxvf jdk-8u311-linux-x64.tar.gz -C /usr/lib/jvm
```

（5）修改环境变量。

使用 sudo vi ~/.zshrc 命令来修改环境变量。

```
shiyanlou:~/ $ sudo vi ~/.zshrc
……
export JAVA_HOME=/usr/lib/jvm/jdk1.8.0_311
export JRE_HOME=$JAVA_HOME/jre
export CLASSPATH=.:$JAVA_HOME/lib:$JAVA_HOME/jre/lib
export PATH=$PATH:$JAVA_HOME/bin:$JRE_HOME
……
```

主要是修改 JAVA_HOME 的路径，其余可以保持不变。

（6）使环境变量马上生效。

使用 source ~/.zshrc 命令使环境变量马上生效。

```
shiyanlou:~/ $ source ~/.zshrc
```

（7）验证是否安装成功。

使用 java -version 命令查看 JDK 的版本。

```
shiyanlou:~/ $ java -version
java version "1.8.0_311"
Java(TM) SE Runtime Environment (build 1.8.0_311-b11)
Java HotSpot(TM) 64-Bit Server VM (build 25.311-b11,mixed mode)
```

4．Hadoop 伪分布式安装部署

（1）下载 Hadoop。

本项目选用的 Hadoop 版本是 2018 年 11 月 19 日发布的稳定版 Hadoop 2.9.2，可以先从 Hadoop 官网下载 hadoop-2.9.2.tar.gz，将其存放在 /home/shiyanlou 中。也可以使用 wget 命令下载蓝桥云课提供的资源（https://labfile.oss.aliyuncs.com/courses/7758/hadoop-

2.9.2.tar.gz）。

```
shiyanlou:~/ $ wget https://labfile.oss.aliyuncs.com/courses/7758/hadoop-2.9.2.tar.gz
```

（2）解压缩并设置环境变量。

使用 tar -zxvf hadoop-2.9.2.tar.gz 命令解压缩。

```
shiyanlou:~/ $ tar -zxvf hadoop-2.9.2.tar.gz
```

设置环境变量。

```
shiyanlou:~/ $ vi ~/.zshrc
……
export HADOOP_HOME=/home/shiyanlou/hadoop-2.9.2
export PATH=$JAVA_HOME/bin:$HADOOP_HOME/bin:$HADOOP_HOME/sbin:$PATH
……
```

使用 source ~/.zshrc 命令使环境变量马上生效。

```
shiyanlou:~/ $ source ~/.zshrc
```

（3）配置 Hadoop。

Hadoop 配置文件有很多，都位于 $HADOOP_HOME/etc/hadoop 下。

下面简单描述一下几个重要的配置文件。

hadoop-env.sh：运行 Hadoop 要用到的环境变量。

core-site.xml：核心配置项，包括 HDFS、MapReduce 和 Yarn 常用的 I/O 设置等。

hdfs-site.xml：HDFS 相关进程的配置项，包括 NameNode、SecondaryNameNode、DataNode 等。

yarn-site.xml：Yarn 相关进程的配置项，包括 ResourceManager、NodeManager 等。

mapred-site.xml：MapReduce 相关进程的配置项。

slaves：从节点配置文件，通常每行 1 个从节点主机名。

log4j.properties：系统日志、NameNode 审计日志、JVM 进程日志的配置项。

Hadoop 的配置文件繁多，我们可采用最小配置，即只修改 6 个配置文件，其余文件保持默认状态。

第 1 个：配置 hadoop-env.sh。

配置如下：

```
shiyanlou:hadoop/ $ vi hadoop-env.sh
export JAVA_HOME=/usr/lib/jvm/java-8-openjdk-amd64
```

第 2 个：配置 core-site.xml。

配置如下：

```
shiyanlou:hadoop/ $ vi core-site.xml
<property>
    <name>fs.defaultFS</name>
    <value>hdfs://localhost:9000</value>
</property>
<property>
    <name>hadoop.tmp.dir</name>
    <value>/home/shiyanlou/hadoop-2.9.2/tmp</value>
</property>
```

配置 fs.defaultFS 指定 Hadoop 所使用的文件系统的 URI（统一资源标识符），示例中的 URI 包含协议（HDFS）、NameNode 的 IP 地址（或者机器名）和端口（9000）。

配置 hadoop.tmp.dir 指定 Hadoop 运行时产生的临时文件的存储目录。

第 3 个：配置 hdfs-site.xml。

配置如下：

```
shiyanlou:hadoop/ $ vi hdfs -site.xml
<property>
    <name>dfs.replication</name>
    <value>1</value>
</property>
<property>
    <name>dfs.secondary.http.address</name>
    <value>localhost:50090</value>
</property>
```

配置 dfs.replication 指定数据副本的数量，由于是伪分布式模式，只有 1 个节点，所以这里设置为 1 即可。

配置 dfs.secondary.http.address 指定 Secondary NameNode 的地址和端口。

第 4 个：配置 mapred-site.xml。

将 mapred-site.xml.template 文件复制一份为 mapred-site.xml 文件。

```
shiyanlou:hadoop/ $ cp mapred-site.xml.templatemapred-site.xml
```

配置如下：

```
shiyanlou:hadoop/ $ vi mapred-site.xml
<property>
    <name>mapreduce.framework.name</name>
    <value>yarn</value>
</property>
```

配置 mapreduce.framework.name 指定 MapReduce 运行在 Yarn 上。

第 5 个：配置 yarn-site.xml。

配置如下：

```
shiyanlou:hadoop/ $ vi yarn-site.xml
<property>
    <name>yarn.resourcemanager.hostname</name>
    <value>localhost</value>
</property>
<property>
    <name>yarn.nodemanager.aux-services</name>
    <value>mapreduce_shuffle</value>
</property>
```

配置 yarn.resourcemanager.hostname 指定 Yarn 的 ResourceManager 的地址。

配置 yarn.nodemanager.aux-services 指定 Shuffle 机制。

第 6 个：配置 slaves。

配置如下：

```
shiyanlou:hadoop/ $ vi slaves
localhost
```

配置集群中的从节点，一行一个机器名（或 IP 地址）。这里因为是伪分布式模式，所以 localhost 既是主节点，又是从节点。

（4）格式化文件系统。

在主节点 master 上执行 hdfs namenode -format 命令格式化 HDFS 文件系统。

```
shiyanlou:hadoop-2.9.2/ $ bin/hdfs namenode -format
```

注意：路径是 /home/shiyanlou/hadoop-2.9.2/bin 下面的 hdfs 命令。

命令执行成功后，会出现 core-site.xml 中 hadoop.tmp.dir 配置项指定的目录。

需要注意的是，只需要进行一次格式化。如果格式化出错，修改相关配置后，需要先把 hadoop.tmp.dir 配置的目录删除，才能再次格式化。

（5）启动 Hadoop 并验证。

首先使用命令 start-dfs.sh 启动 HDFS。

```
shiyanlou:hadoop-2.9.2/ $ sbin/start-dfs.sh
Starting namenodes on [localhost]
localhost:starting namenode,logging to /home/shiyanlou/hadoop-2.9.2/logs/
hadoop-shiyanlou-namenode-12a696ffc891.out
```

```
localhost:starting datanode,logging to /home/shiyanlou/hadoop-2.9.2/logs/
hadoop-shiyanlou-datanode-12a696ffc891.out
  Starting secondary namenodes [localhost]
localhost:starting secondarynamenode,logging to /home/shiyanlou/hadoop-2.9.2/
logs/hadoop-shiyanlou-secondarynamenode-12a696ffc891.out
```

然后使用 jps 命令查看启动进程。

```
shiyanlou:hadoop-2.9.2/ $ jps
656 NameNode
787 DataNode
1078 Jps
954 SecondaryNameNode
```

通过输出可以看到 NameNode、DataNode 和 SecondaryNameNode 这 3 个进程已经启动了，而且提示了对应的日志文件（*_.out 就是日志文件）。如果启动失败，可以去日志文件查看报错信息。

再使用 start-yarn.sh 命令启动 Yarn。

```
shiyanlou:hadoop-2.9.2/ $ sbin/start-yarn.sh
starting yarn daemons
starting resourcemanager,logging to/home/shiyanlou/hadoop-2.9.2/logs/yarn-
shiyanlou-resourcemanager-12a696ffc891.out
localhost:starting nodemanager,logging to/home/shiyanlou/hadoop-2.9.2/logs/
yarn-shiyanlou-nodemanager-12a696ffc891.out
```

接着使用 jps 命令查看启动进程。

```
shiyanlou:hadoop-2.9.2/ $ jps
4626 NameNode
5171 ResourceManager
4757 DataNode
5592 Jps
5275 NodeManager
4924 SecondaryNameNode
```

通过输出可以看到 ResourceManager 和 NodeManager 这 2 个进程已经启动了。

下面我们确认一下已经启动的 Java 进程。

```
shiyanlou:~/ $ jps
2465 NodeManager
2211 SecondaryNameNode
2517 Jps
1912 NameNode
2012 DataNode
2366 ResourceManager
```

可以看到除了 Jps 进程本身外，启动了 5 个 Java 进程。

Hadoop 也提供了基于 Web 浏览器的管理工具，可以用来验证 Hadoop 集群是否部署成功且正确启动。其中 HDFS Web UI 的默认地址为 http://NameNodeIP:50070，这里为 http://localhost:50070，运行界面如图附 1.1 所示。

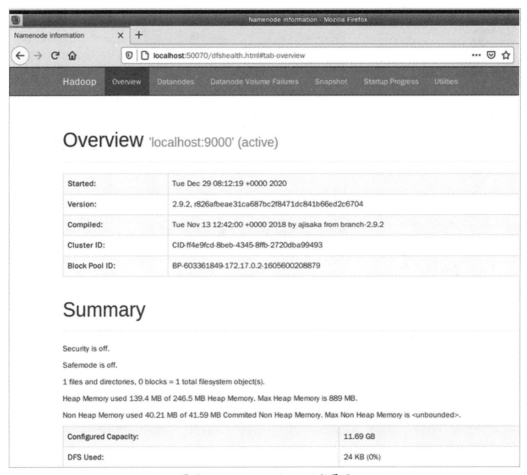

图附 1.1　HDFS Web UI 运行界面

Yarn Web UI 的默认地址为 http://ResourceManagerIP:8088，这里为 http://localhost:8088，运行界面如图附 1.2 所示。

图附 1.2 Yarn Web UI 运行界面

（6）关闭 Hadoop。

关闭 Hadoop 集群时，与启动顺序正好相反，先关闭 Yarn，再关闭 HDFS。

关闭 Yarn：

```
shiyanlou:hadoop-2.9.2/ $ sbin/stop-yarn.sh
stopping yarn daemons
stopping resourcemanager
localhost:stopping nodemanager
localhost:nodemanager did not stop gracefully after 5 seconds:killing with kill -9
no proxyserver to stop
```

关闭 HDFS：

```
shiyanlou:hadoop-2.9.2/ $ sbin/stop-dfs.sh
Stopping namenodes on [localhost]
localhost:stopping namenode
localhost:stopping datanode
Stopping secondary namenodes [localhost]
localhost:stopping secondarynamenode
```

注意：还可以使用 start-all.sh 命令一起开启 HDFS 和 Yarn，使用 stop-all.sh 命令一起关闭 Yarn 和 HDFS。

附录2 掌握 HDFS Shell 操作

这里帮助读者熟悉 HDFS Shell 文件系统的各种操作命令的使用，包括文件的创建、修改、删除、权限修改等，文件夹的创建、复制、删除、重命名等。

HDFS Shell 命令格式类似于 Linux 终端对文件的操作，如 ls、mkdir、rm 等。HDFS 文件系统命令的入口是 hadoop fs，其语法是 hadoop fs [generic options]。hdfs dfs [generic options] 命令也可以使用，它们两者的区别在于：hadoop fs 命令使用面更广，可以操作任何文件系统，比如本地文件、HDFS 文件、HFTP 文件、S3 文件系统等；而 hdfs dfs 命令则是针对 HDFS 文件系统的操作。

下面将对 hadoop fs 中常用的命令选项做详细说明。开始操作前，先启动 Hadoop 的 HDFS。

```
shiyanlou:~/ $ start-dfs.sh
......
```

1. ls 和 ls -R 命令

命令：ls。

功能：显示文件的元数据信息或者目录包含的文件列表信息。信息从左往右依次为权限、副本数、用户 ID、用户组 ID、文件（目录）大小（字节）、修改时间、目录。

格式：hadoop fs -ls [path]。

示例：显示 HDFS 根目录 / 下的所有文件与目录的基本信息。

```
shiyanlou:~/ $ vi a.txt
shiyanlou:~/ $ hadoop fs -put a.txt /
shiyanlou:~/ $ hadoop fs -ls /
Found 1 items
-rw-r--r--   1 shiyanlou supergroup          4 2022-03-03 12:43 /a.txt
```

解释："vi a.txt" 的作用是创建一个 a.txt 文件，"hadoop fs -put a.txt /" 的作用是把当前目录下的 a.txt 文件上传到 HDFS 的根目录 / 下。"hadoop fs -ls /" 的作用是显示 HDFS 根目录 / 下的所有文件与目录的基本信息，和 Linux 中的 ls 命令类似。

命令：ls -R。

功能：ls 命令的递归版本，类似 Linux 中的 ls -R 命令。

格式：hadoop fs -ls -R [path]。

示例：递归地显示根目录 / 下的所有文件夹和文件。

```
shiyanlou:~/ $ hadoop fs -mkdir /a
shiyanlou:~/ $ hadoop fs -put a.txt /a/b.txt
shiyanlou:~/ $ hadoop fs -put a.txt /a/c.txt
shiyanlou:~/ $ hadoop fs -ls -R /
drwxr-xr-x   - shiyanlou supergroup          0 2022-03-03 12:50 /a
-rw-r--r--   1 shiyanlou supergroup          4 2022-03-03 12:50 /a/b.txt
-rw-r--r--   1 shiyanlou supergroup          4 2022-03-03 12:50 /a/c.txt
-rw-r--r--   1 shiyanlou supergroup          4 2022-03-03 12:43 /a.txt
```

解释:"hadoop fs -mkdir /a"的作用是在 HDFS 根目录下创建一个子目录 a。"hadoop fs -put a.txt /a/b.txt"的作用是把当前目录下的 a.txt 文件上传到 HDFS 中的目录 /a 下,并命名为 b.txt。"hadoop fs -put a.txt /a/c.txt"的作用是把当前目录下的 a.txt 文件上传到 HDFS 中的目录 /a 下,并命名为 c.txt。"hadoop fs -ls -R /"的作用是递归地显示 HDFS 根目录 / 下的所有文件和子目录的信息。

2. du 和 du -s 命令

命令:du。

功能:显示文件的大小或者目录中包含的所有文件的大小。

格式:hadoop fs -du [path]。

示例:显示目录 /a 中每个文件的大小。

```
shiyanlou:~/ $ hadoop fs -du /a
4   /a/b.txt
4   /a/c.txt
shiyanlou:~/ $ hadoop fs -du /a.txt
4   /a.txt
```

解释:"hadoop fs -du /a"的作用是显示目录 /a 下每个文件的大小,默认单位是字节。"hadoop fs -du /a.txt"的作用是显示文件 /a.txt 的大小。

命令:du -s。

功能:显示目录下所有文件大小之和,单位是字节。

格式:hadoop fs -du -s [path]。

示例:显示目录 /a 下所有文件大小之和。

```
shiyanlou:~/ $ hadoop fs -du -s /a
8   /a
```

解释:"hadoop fs -du -s /a"的作用是显示目录 /a 下所有文件大小之和,这里有 2 个文件,每个文件 4 字节,所以和为 8 字节。

3. count 和 mkdir 命令

命令：count。

功能：统计文件（目录）的数量和文件（目录）的大小信息。

格式：hadoop fs –count [path]。

示例：显示 /a 目录下包含的文件数量。

```
shiyanlou:~/ $ hadoop fs -count /a
1       2       8       /a
```

解释："hadoop fs –count /a"的作用是显示 /a 目录下包含的文件数量，其中 1 表示 1 个文件，2 表示含有 2 个文件，8 表示 2 个文件的大小总和为 8 字节。

命令：mkdir。

功能：创建指定的一个或多个目录，–p 选项用于递归创建子目录。类似于 Linux 的 mkdir –p。

格式：hadoop fs –mkdir [–p] [paths]。

示例：在 HDFS 目录 / 下创建目录 b。

```
shiyanlou:~/ $ hadoop fs -mkdir /b
shiyanlou:~/ $ hadoop fs -ls /
Found 3 items
drwxr-xr-x   - shiyanlou supergroup          0 2022-03-03 12:50 /a
-rw-r--r--   1 shiyanlou supergroup          4 2022-03-03 12:43 /a.txt
drwxr-xr-x   - shiyanlou supergroup          0 2022-03-03 13:30 /b
```

解释："hadoop fs –mkdir /b"的作用是在 HDFS 目录 / 下创建目录 b。"hadoop fs –ls /"的作用是查看 / 目录下是否有 b 目录。

4. mv 和 cp 命令

命令：mv。

功能：移动 HDFS 的文件到指定位置，第一个参数表示被移动文件位置，第二个参数表示移动的目标位置（通常为目录）。

格式：hadoop fs –mv [src] [dst]。

示例：将 /a 目录下的文件 b.txt 移动到 / 目录下。

```
shiyanlou:~/ $ hadoop fs -mv /a/b.txt /
shiyanlou:~/ $ hadoop fs -ls -R /
drwxr-xr-x   - shiyanlou supergroup          0 2022-03-03 13:35 /a
-rw-r--r--   1 shiyanlou supergroup          4 2022-03-03 12:50 /a/c.txt
```

```
-rw-r--r--   1 shiyanlou supergroup          4 2022-03-03 12:43 /a.txt
drwxr-xr-x   - shiyanlou supergroup          0 2022-03-03 13:30 /b
-rw-r--r--   1 shiyanlou supergroup          4 2022-03-03 12:50 /b.txt
```

解释："hadoop fs -mv /a/b.txt /"的作用是将目录 /a 下的 b.txt 文件移动到目录 / 下。"hadoop fs -ls -R /"的作用是递归地查看目录 / 下的所有文件及子目录，用来查看操作是否正确。

命令：cp。

功能：将文件从源路径复制到目标路径。源路径是多个文件时，目标路径必须是文件目录。

格式：hadoop fs -cp [src] [dst]。

示例：将文件 /a.txt 复制到 /a 目录下。

```
shiyanlou:~/ $ hadoop fs -cp /a.txt /a/
shiyanlou:~/ $ hadoop fs -ls -R /
drwxr-xr-x   - shiyanlou supergroup          0 2022-03-03 13:40 /a
-rw-r--r--   1 shiyanlou supergroup          4 2022-03-03 13:40 /a/a.txt
-rw-r--r--   1 shiyanlou supergroup          4 2022-03-03 12:50 /a/c.txt
-rw-r--r--   1 shiyanlou supergroup          4 2022-03-03 12:43 /a.txt
drwxr-xr-x   - shiyanlou supergroup          0 2022-03-03 13:30 /b
-rw-r--r--   1 shiyanlou supergroup          4 2022-03-03 12:50 /b.txt
```

解释："hadoop fs -cp /a.txt /a/"的作用是将文件 /a.txt 复制到 /a 目录下。"hadoop fs -ls -R /"的作用是递归地查看目录 / 下的所有文件及子目录，用来查看操作是否正确。

5. rm 和 rm -r 命令

命令：rm。

功能：删除文件或者空文件目录。

注意：如果删除非空文件目录，操作会失败。

格式：hadoop fs -rm [path]。

示例：删除文件 /a/a.txt。

```
shiyanlou:~/ $ hadoop fs -rm /a/a.txt
Deleted /a/a.txt
shiyanlou:~/ $ hadoop fs -ls /a
Found 1 items
-rw-r--r--   1 shiyanlou supergroup          4 2022-03-03 12:50 /a/c.txt
```

解释："hadoop fs -rm /a/a.txt"的作用是删除文件 /a/a.txt。"hadoop fs -ls /a"的作用是查看目录 /a 下的所有文件，用来查看操作是否正确。

命令：rm –r。

功能：递归删除文件或者文件夹，文件夹可以包含子目录。

格式：hadoop fs –rm –r [path]。

示例：递归删除 /a 目录及其下所有文件和文件夹。

```
shiyanlou:~/ $ hadoop fs -rm -r /a
Deleted /a
shiyanlou:~/ $ hadoop fs -ls -R /
-rw-r--r--   1 shiyanlou supergroup          4 2022-03-03 12:43 /a.txt
drwxr-xr-x   - shiyanlou supergroup          0 2022-03-03 13:30 /b
-rw-r--r--   1 shiyanlou supergroup          4 2022-03-03 12:50 /b.txt
```

解释："hadoop fs –rm –r /a"的作用是删除目录 /a。"hadoop fs –ls –R /"的作用是递归地查看目录 / 下的所有文件及子目录的操作是否正确。

6．cat 和 text 命令

命令：cat。

功能：将指定文件内容输出到标准输出 stdout。

格式：hadoop fs –cat [path]。

示例：输出 HDFS 目录 / 下的 a.txt 文件到标准输出。

```
shiyanlou:~/ $ hadoop fs -cat /a.txt
123
```

解释："hadoop fs –cat /a.txt"的作用是输出 HDFS 目录 / 下的 a.txt 文件到标准输出。

命令：text。

功能：和 cat 一样。

7．put 和 getmerge 命令

命令：put。

功能：从本地文件系统复制单个或多个源路径到目标文件系统。

格式：hadoop fs –put [localsrc1] [localsrc2] …… [dst]。

示例：将 Linux 当前目录下的文件 a.txt 复制到 HDFS 目录 /a 下面，并重命名为 b.txt。

```
shiyanlou:~/ $ hadoop fs -mkdir /a
shiyanlou:~/ $ hadoop fs -put a.txt /a/b.txt
shiyanlou:~/ $ hadoop fs -ls /a
Found 1 items
-rw-r--r--   1 shiyanlou supergroup          4 2022-03-03 13:57 /a/b.txt
```

解释:"hadoop fs –mkdir /a"的作用是在 HDFS 中创建目录 /a。"hadoop fs –put a.txt /a/b.txt"的作用是将 Linux 当前目录下的 a.txt 文件复制到 HDFS 目录 /a 下面,并重命名为 b.txt。"hadoop fs –ls /a"的作用是查看目录 /a 下的所有文件的操作是否正确。

命令:getmerge。

功能:将 HDFS 源目录中所有文件合并,并下载到本地,即 merge 和 get 两个动作。在合并时可以在每个文件结束处添加换行(增加参数 –nl)。

格式:hadoop fs –getmerge [–nl] [src] [localdst]。

示例:将 HDFS 目录 /a 下的文件合并,输出到本地文件系统,保存为 a.txt。

```
shiyanlou:~/ $ hadoop fs -cp /a.txt /a/
shiyanlou:~/ $ hadoop fs -ls /a
Found 2 items
-rw-r--r--   1 shiyanlou supergroup          4 2022-03-03 14:12 /a/a.txt
-rw-r--r--   1 shiyanlou supergroup          4 2022-03-03 13:57 /a/b.txt
shiyanlou:~/ $ hadoop fs -getmerge /a a.txt
shiyanlou:~/ $ cat a.txt
123
123
```

解释:"hadoop fs –cp /a.txt /a/"的作用是将 a.txt 文件复制一份到目录 /a 中。"hadoop fs –ls /a"的作用是查看目录 /a 下有 2 个文件。"hadoop fs –getmerge /a a.txt"的作用是将 HDFS 目录 /a 下的文件合并,输出到本地文件系统,保存为 a.txt。"cat a.txt"的作用是查看文件内容,可以看到 2 个文件的内容合并到 1 个文件中了。

8. chmod、chown 和 chgrp 命令

命令:chmod。

功能:修改文件权限。使用 –R 选项可以在目录结构下递归更改文件权限。

注意:该命令的使用者必须是文件的所有者或者超级用户。

格式:hadoop fs –chmod [–R] 权限 [path]。

示例:更改 /a/a.txt 文件的权限为 766。

```
shiyanlou:~/ $ hadoop fs -chmod 766 /a/a.txt
shiyanlou:~/ $ hadoop fs -ls /a
Found 2 items
-rwxrw-rw-   1 shiyanlou supergroup          4 2022-03-03 14:12 /a/a.txt
-rw-r--r--   1 shiyanlou supergroup          4 2022-03-03 13:57 /a/b.txt
```

解释:"hadoop fs –chmod 766 /a/a.txt"的作用是更改 /a/a.txt 文件的权限为 766。"hadoop fs –ls /a"的作用是查看 /a/a.txt 文件的权限是否已经改变。

命令：chown。

功能：更改文件的所有者。–R 选项表示递归更改所有子目录。

注意：该命令的使用者必须是文件所有者或者超级用户。

格式：hadoop fs –chown [–R] [OWNER] [:[GROUP]] [path]。

示例：更改 /a/a.txt 文件的所有者为 user1。

```
shiyanlou:~/ $ hadoop fs -chown user1 /a/a.txt
shiyanlou:~/ $ hadoop fs -ls /a
Found 2 items
-rwxrw-rw-   1 user1     supergroup      4 2022-03-03 14:12 /a/a.txt
-rw-r--r--   1 shiyanlou supergroup      4 2022-03-03 13:57 /a/b.txt
```

解释："hadoop fs –chown user1 /a/a.txt"的作用是更改 /a/a.txt 文件的所有者为 user1。"hadoop fs –ls /a"的作用是查看 /a/a.txt 文件的所有者是否已经改变。

命令：chgrp。

功能：更改文件所属的用户组。–R 选项表示递归处理子目录。注意：该命令的使用者必须是文件所有者或者超级用户。

格式：hadoop fs –chgrp [–R] GROUP [path]。

示例：更改 /a/a.txt 的组为 group1。

```
shiyanlou:~/ $ hadoop fs -chgrp group1 /a/a.txt
shiyanlou:~/ $ hadoop fs -ls /a
Found 2 items
-rwxrw-rw-   1 user1     group1          4 2022-03-03 14:12 /a/a.txt
-rw-r--r--   1 shiyanlou supergroup      4 2022-03-03 13:57 /a/b.txt
```

解释："hadoop fs –chgrp group1 /a/a.txt"的作用是更改 /a/a.txt 文件所属的用户组为 group1。"hadoop fs –ls /a"的作用是查看 /a/a.txt 文件所属的用户组是否已经改变。

9．setrep 和 expunge 命令

命令：setrep。

功能：改变文件的副本数。–R 选项用于递归改变目录下所有文件的副本数。此后，HDFS 会根据该参数自动进行复制或清除工作。参数 –w 表示等副本操作结束后再退出命令。

格式：hadoop fs –setrep [–R] [–w] [rep] [path]。

示例：设置文件 /a/a.txt 的副本数为 2。

```
shiyanlou:~/ $ hadoop fs -setrep 2 /a/a.txt
Replication 2 set:/a/a.txt
shiyanlou:~/ $ hadoop fs -ls /a
Found 2 items
-rwxrw-rw-   2 user1      group1              4 2022-03-03 14:12 /a/a.txt
-rw-r--r--   1 shiyanlou  supergroup          4 2022-03-03 13:57 /a/b.txt
```

解释:"hadoop fs –setrep 2 /a/a.txt"的作用是设置文件 /a/a.txt 的副本数为 2。"hadoop fs –ls /a"的作用是查看 /a/a.txt 文件的副本数是否已经改变。

命令:expunge。

功能:清空回收站。

注意:清空后数据将不可恢复。

格式:hadoop fs –expunge。

示例:清空回收站。

```
shiyanlou:~/ $ hadoop fs -expunge
```

附录 3　通过 WordCount 熟悉 MapReduce

启动 HDFS 和 Yarn。

```
shiyanlou:~/ $ start-all.sh
This script is Deprecated. Instead use start-dfs.sh and start-yarn.sh
Starting namenodes on [localhost]
localhost:starting namenode,logging to /opt/hadoop-2.9.2/logs/hadoop-shiyanlou-namenode-a557ab4e72a5.out
localhost:starting datanode,logging to /opt/hadoop-2.9.2/logs/hadoop-shiyanlou-datanode-a557ab4e72a5.out
Starting secondary namenodes [0.0.0.0]
The authenticity of host '0.0.0.0 (0.0.0.0)' can't be established.
ECDSA key fingerprint is SHA256:K4O0OUGh2ZloSwzKI7L7fcYCdBuUbJORA12+UsMaUAI.
Are you sure you want to continue connecting (yes/no)? yes
0.0.0.0:Warning:Permanently added '0.0.0.0' (ECDSA) to the list of known hosts.
0.0.0.0:starting secondarynamenode,logging to /opt/hadoop-2.9.2/logs/hadoop-shiyanlou-secondarynamenode-a557ab4e72a5.out
starting yarn daemons
starting resourcemanager,logging to /opt/hadoop-2.9.2/logs/yarn-shiyanlou-resourcemanager-a557ab4e72a5.out
```

```
localhost:starting nodemanager,logging to /opt/hadoop-2.9.2/logs/yarn-
shiyanlou-nodemanager-a557ab4e72a5.out
```

使用 Eclipse 作为 IDE（集成开发环境），为了方便管理 jar 包，创建一个 Maven 项目，添加以下依赖。

```xml
<dependencies>
    <dependency>
        <groupId>junit</groupId>
        <artifactId>junit</artifactId>
        <version>4.12</version>
    </dependency>
    <dependency>
        <groupId>log4j</groupId>
        <artifactId>log4j</artifactId>
        <version>1.2.17</version>
    </dependency>
    <dependency>
        <groupId>org.apache.hadoop</groupId>
        <artifactId>hadoop-client</artifactId>
        <version>2.9.2</version>
    </dependency>
    <dependency>
        <groupId>org.apache.hadoop</groupId>
        <artifactId>hadoop-common</artifactId>
        <version>2.9.2</version>
    </dependency></dependencies>
```

WordCount 的完整代码如程序清单 5.1 所示。

程序清单 5.1

```java
package org.lanqiao.bigdata;

import java.net.URI;

import org.apache.hadoop.conf.Configuration;
import org.apache.hadoop.fs.FileSystem;
import org.apache.hadoop.fs.Path;
import org.apache.hadoop.io.LongWritable;
import org.apache.hadoop.io.Text;
import org.apache.hadoop.mapreduce.Job;
import org.apache.hadoop.mapreduce.Mapper;
```

```java
import org.apache.hadoop.mapreduce.Reducer;
import org.apache.hadoop.mapreduce.lib.input.FileInputFormat;
import org.apache.hadoop.mapreduce.lib.output.FileOutputFormat;

public class WordCount {
    public static void main(String[] args) throws Exception {
        Configuration conf=new Configuration();
        // 获取集群访问路径
        FileSystem fileSystem=FileSystem.get(new URI("hdfs://localhost:9000/"),conf);
        // 判断输出路径是否已经存在,如果存在则删除
        Path outPath=new Path(args[1]);
        if(fileSystem.exists(outPath)){
            fileSystem.delete(outPath,true);
        }
        // 生成 job,并指定 job 的名称
        Job job=Job.getInstance(conf,"word-count");
        // 指定打成 jar 后的运行类
        job.setJarByClass(WordCount.class);
        // 指定 mapper 类
        job.setMapperClass(MyMapper.class);
        // 指定 reducer 类
        job.setReducerClass(MyReducer.class);
        // 指定 mapper 的输出类型
        job.setMapOutputKeyClass(Text.class);
        job.setMapOutputValueClass(LongWritable.class);
        // 指定 reducer 的输出类型
        job.setOutputKeyClass(Text.class);
        job.setOutputValueClass(LongWritable.class);

        FileInputFormat.addInputPath(job,new Path(args[0]));
        FileOutputFormat.setOutputPath(job,new Path(args[1]));
        System.exit(job.waitForCompletion(true)?0:1);
    }

    /**
     * Mapper<LongWritable,Text,Text,LongWritable> 中前两个参数表示 map 函数处理前的
     输入数据的 key 和 value 的数据类型
     * 后两个参数表示 map 函数处理后的输出数据的 key 和 value 的数据类型
     * 这里使用的数据类型都必须是下节介绍的 MapReduce 中的数据类型
     */
    static class MyMapper extends Mapper<LongWritable,Text,Text,LongWritable>{
```

```java
            /**
             * @param k1 表示每行偏移量
             * @param v1 表示每行行内容
             * @param context 表示当前上下文环境对象
             */
            protected void map(LongWritable k1,Text v1,Context context) throws java.io.IOException ,InterruptedException {
                    // 按照空格切割每一行的内容,每一个单词构建一个键值对,如"hello-1"
                    final String[] splited=v1.toString().split("");
                    for (String word:splited) {
                            // 使用context上下文对象的write方法把该键值对发送到下一个环节——排序
                                    context.write(new Text(word),new LongWritable(1));
                    }
            };
    }


    /**
     * Reducer<Text,LongWritable,Text,LongWritable> 中前两个参数表示 reduce 函数处理前的输入数据的 key 和 value 的数据类型
     * 后两个参数表示 reduce 函数处理后的输出数据的 key 和 value 的数据类型
     */
    static class MyReducer extends Reducer<Text,LongWritable,Text,LongWritable>{
            /**
             * @param k2 表示单词
             * @param v2s 表示单词数量的集合
             * @paramctx 表示当前上下文环境对象
             */
            protected void reduce(Text k2,java.lang.Iterable<LongWritable> v2s,Context ctx) throws java.io.IOException ,InterruptedException {
                    long times=0L;
                    // 针对每一个单词累计求和
                    for (LongWritable count:v2s) {
                            times+=count.get();
                    }
                    // 把每个单词的最终统计结果发送给下一个环节——输出
                    ctx.write(k2,new LongWritable(times));
            };
    }
}
```

下面我们来准备数据文件。

```
shiyanlou:~/ $ cat file1.txt
Dear Bear River
Car Car River
Dear Car Bear
shiyanlou:~/ $ cat file2.txt
Dear River River
Dear Car
Car Car Bear
shiyanlou:~/ $ cat file3.txt
Dear Bear Dear
River River Bear
Car Bear River
```

把这 3 个文件上传到 HDFS 中的 /input 目录下，如果该目录下已经有文件了，记得先清空该目录。

```
shiyanlou:~/ $ hadoop fs -mkdir /input
shiyanlou:~/ $ hadoop fs -put file1.txt /input/
shiyanlou:~/ $ hadoop fs -put file2.txt /input/
shiyanlou:~/ $ hadoop fs -put file3.txt /input/
```

在 Eclipse 中把已经完成的项目代码打成包 wc.jar，再执行该 jar 包，命令如下。

```
shiyanlou:~/ $ hadoop jar wc.jar org.lanqiao.bigdata.WordCount /input /out
```

其中 hadoop 是命令的名称，jar 是执行的文件类型，org.lanqiao.bigdata.WordCount 为可执行类的包路径，/input 为输入目录，/out 为输出目录，输出的结果如下所示。

```
shiyanlou:~/ $ hadoop fs -cat /out/part-r-00000
Bear    6
Car     7
Dear    6
River   7
```

可以看到，结果正确。

附录 4　深入理解 MapReduce

1．Shuffle 分区

如果我们希望采用 WordCount 时，首字母为 A~M 的单词属于一个区，首字母为 N~Z 的单词属于另一个区，那么怎么实现自定义分区呢？

首先启动 HDFS 和 Yarn。

```
shiyanlou:~/ $ start-all.sh
……
```

然后使用 Eclipse 作为 IDE，保持开发环境和之前的实验环境一样就可以了，完整代码如程序清单 5.2 所示。

程序清单 5.2

```java
package org.lanqiao.bigdata;

import java.net.URI;

import org.apache.hadoop.conf.Configuration;
import org.apache.hadoop.fs.FileSystem;
import org.apache.hadoop.fs.Path;
import org.apache.hadoop.io.LongWritable;
import org.apache.hadoop.io.Text;
import org.apache.hadoop.mapreduce.Job;
import org.apache.hadoop.mapreduce.Mapper;
import org.apache.hadoop.mapreduce.Reducer;
import org.apache.hadoop.mapreduce.lib.input.FileInputFormat;
import org.apache.hadoop.mapreduce.lib.output.FileOutputFormat;
import org.apache.hadoop.mapreduce.lib.partition.HashPartitioner;

public class CustomPartition {
    public static void main(String[] args) throws Exception {
        Configuration conf=new Configuration();
        // 获取集群访问路径
        FileSystem fileSystem=FileSystem.get(new URI("hdfs://localhost:9000/"),conf);
        // 判断输出路径是否已经存在，如果存在则删除
        Path outPath=new Path(args[1]);
        if(fileSystem.exists(outPath)){
            fileSystem.delete(outPath,true);
        }
        // 生成 job，并指定 job 的名称
        Job job=Job.getInstance(conf,"word-count");
        // 指定打成 jar 后的运行类
        job.setJarByClass(CustomPartition.class);
        // 指定 mapper 类
        job.setMapperClass(MyMapper.class);
        // 指定 reducer 类
```

```java
        job.setReducerClass(MyReducer.class);
            // 指定mapper的输出类型
        job.setMapOutputKeyClass(Text.class);
        job.setMapOutputValueClass(LongWritable.class);
            // 指定reducer的输出类型
        job.setOutputKeyClass(Text.class);
        job.setOutputValueClass(LongWritable.class);

            // 分区
        job.setPartitionerClass(MyPartitioner.class);
            // 同时指定分区数量
        job.setNumReduceTasks(2);

        FileInputFormat.addInputPath(job,new Path(args[0]));
        FileOutputFormat.setOutputPath(job,new Path(args[1]));
        System.exit(job.waitForCompletion(true)?0:1);
    }

    /**
     * 自定义分区类。分区规则：首字母为[A,M]的单词为一个分区，首字母为[N,Z]的单词为另一个分区
     * key和value是Map的输出键值对，numReduceTasks是分区数
     */
    static class MyPartitioner extends HashPartitioner<Text,LongWritable>{
        public int getPartition(Text key,LongWritable value,int numReduceTasks) {
            char k=key.toString().charAt(0);// 首字母
            if(k>='a'&&k<='m'||k>='A'&&k<='M'){
                return 0;
            }else {
                return 1;
            }
        }
    }

    static class MyMapper extends Mapper<LongWritable,Text,Text,LongWritable>{
        protected void map(LongWritable k1,Text v1,Context context) throws java.io.IOException ,InterruptedException {
            final String[] splited=v1.toString().split("");
            for (String word:splited) {
```

```
            context.write(new Text(word),new LongWritable(1));
                }
            };
        }


        static class MyReducer extends Reducer<Text,LongWritable,Text,LongWritable>{
            protected void reduce(Text k2,java.lang.Iterable<LongWritable> v2s,Context ctx) throws java.io.IOException ,InterruptedException {
                long times=0L;
                for (LongWritable count:v2s) {
                    times+=count.get();
                }
        ctx.write(k2,new LongWritable(times));
            };
        }
}
```

准备好数据文件 a.txt。

```
shiyanlou:~/ $ cat a.txt
i love you What
you love me
What What
```

将其上传到 HDFS 的 /input 目录下，上传前最好确保该目录是空的。

```
shiyanlou:~/ $ hadoop fs -mkdir /input
shiyanlou:~/ $ hadoop fs -put a.txt /input/
```

把代码打成 jar 包，放到 ~ 目录下，并执行。

```
shiyanlou:~/ $ hadoop jar CustomPartition.jar org.lanqiao.bigdata.CustomPartition /input /out
```

执行完毕后，查看输出结果，得到两个分区文件 part-r-00000 和 part-r-00001，内容符合分区标准。

```
shiyanlou:~/ $ hadoop fs -ls /out
Found 3 items
-rw-r--r--   1   shiyanlou   supergroup   0    2021-09-16 05:31 /out/_SUCCESS
-rw-r--r--   1   shiyanlou   supergroup   16   2021-09-16 05:31 /out/part-r-00000
-rw-r--r--   1   shiyanlou   supergroup   13   2021-09-16 05:31 /out/part-r-00001
```

```
shiyanlou:~/ $ hadoop fs -cat /out/part-r-00000
i       1
love    2
me      1
shiyanlou:~/ $ hadoop fs -cat /out/part-r-00001
What    3
you     2
```

2．Shuffle 排序

下面我们举一个例子，来实现一个复合排序的功能。

首先启动 HDFS 和 Yarn。

```
shiyanlou:~/ $ start-all.sh
……
```

然后准备数据文件 a.txt，注意分隔符为 \t，而不是空格。

```
shiyanlou:~/ $ cat a.txt
zs  20  100
ls  21  95
ww  22  90
zl  21  99
sq  19  88
qb  20  96
```

把该文件上传到 HDFS 上的 /input 目录下。

```
shiyanlou:~/ $ hadoop fs -mkdir /input
shiyanlou:~/ $ hadoop fs -put a.txt /input/
```

该文件的第一列是姓名，第二列是年龄，第三列是成绩。现在我们希望把这个文件的数据先按照年龄升序排列，年龄相同的再按照成绩降序排列，输出到 1 个文件中，形式不变。

使用 Eclipse 作为 IDE，保持开发环境和之前的实验环境一样就可以了，详细代码如程序清单 5.3 所示。

程序清单 5.3

```
package org.lanqiao.bigdata;

import java.io.DataInput;
import java.io.DataOutput;
import java.io.IOException;
import java.net.URI;
```

```java
import org.apache.hadoop.conf.Configuration;
import org.apache.hadoop.fs.FileSystem;
import org.apache.hadoop.fs.Path;
import org.apache.hadoop.io.LongWritable;
import org.apache.hadoop.io.NullWritable;
import org.apache.hadoop.io.Text;
import org.apache.hadoop.io.WritableComparable;
import org.apache.hadoop.mapreduce.Job;
import org.apache.hadoop.mapreduce.Mapper;
import org.apache.hadoop.mapreduce.Reducer;
import org.apache.hadoop.mapreduce.lib.input.FileInputFormat;
import org.apache.hadoop.mapreduce.lib.output.FileOutputFormat;

public class CustomSort {
    public static void main(String[] args) throws Exception {
        Configuration conf=new Configuration();
        // 获取集群访问路径
        FileSystem fileSystem=FileSystem.get(new URI("hdfs://localhost:9000/"),conf);
        // 判断输出路径是否已经存在，如果存在则删除
        Path outPath=new Path(args[1]);
        if(fileSystem.exists(outPath)){
            fileSystem.delete(outPath,true);
        }
        // 生成job，并指定job的名称
        Job job=Job.getInstance(conf,"custom-sort");
        // 指定打成jar后的运行类
        job.setJarByClass(CustomSort.class);
        // 指定mapper类
        job.setMapperClass(MyMapper.class);
        // 指定reducer类
        job.setReducerClass(MyReducer.class);
        // 指定mapper的输出类型
        job.setMapOutputKeyClass(Student.class);
        job.setMapOutputValueClass(NullWritable.class);
        // 指定reducer的输出类型
        job.setOutputKeyClass(Student.class);
        job.setOutputValueClass(NullWritable.class);

        FileInputFormat.addInputPath(job,new Path(args[0]));
        FileOutputFormat.setOutputPath(job,new Path(args[1]));
```

```java
        System.exit(job.waitForCompletion(true)?0:1);
    }

    /**
     * map 函数的输出 key 为 Student 类型，便于使用其重写的 compareTo 方法来排序，value 值我们不关心，所以使用 NullWritable
     */
    static class MyMapper extends Mapper<LongWritable,Text,Student,NullWritable>{
        protected void map(LongWritable k1,Text v1,Context context) throws java.io.IOException ,InterruptedException {
            final String[] splited=v1.toString().split("\t");
            // 拼凑出 Student 类型，作为 key
            final Student s=new Student(splited[0],Integer.parseInt(splited[1]),Integer.parseInt(splited[2]));
            context.write(s,NullWritable.get());
        };
    }

    /**
     * reduce 函数的输出 key 为 Student 类型，便于使用其重写的 toString 方法来输出对象结果到文件中，value 值我们不关心，所以使用 NullWritable
     */
    static class MyReducer extends Reducer<Text,LongWritable,Student,NullWritable>{
        protected void reduce(Student k2,java.lang.Iterable<NullWritable> v2s,Context context) throws java.io.IOException ,InterruptedException {
            // 直接把 Student 对象输出
            context.write(k2,NullWritable.get());
        };
    }

    /**
     * 该类实现了排序规则，实现了 WritableComparable 接口
     */
    static class Student implements WritableComparable<Student>{
        String name;
        int age;
        int score;

        public String getName() {
```

```java
        return name;
    }

    public void setName(String name) {
        this.name=name;
    }

    public int getAge() {
        return age;
    }

    public void setAge(int age) {
        this.age=age;
    }

    public int getScore() {
        return score;
    }

    public void setScore(int score) {
    this.score=score;
    }

    public Student(){}

    public Student(String name,int age,int score){
        this.name=name;
        this.age=age;
        this.score=score;
    }

    /**
     * 下面的两个方法主要用来实现 MapReduce 中的序列化和反序列化
     * 因为 Student 对象要作为 key 在 Hadoop 节点之间进行数据传输
     * 反序列化
     */
    public void readFields(DataInput in) throws IOException {
        this.name=in.readUTF();
        this.age=in.readInt();
        this.score=in.readInt();
    }
```

```java
/**
 * 序列化
 */
public void write(DataOutput out) throws IOException {
    out.writeUTF(name);
    out.writeInt(age);
    out.writeInt(score);
}

/**
 * 当key是Student类型时，默认按照compareTo()方式进行比较
 * 在比较两个Student对象时，先按照age升序排列，再按照score降序排列
 */
public int compareTo(Student o) {
    if(this.age<o.age) {
        return -1;
    }else if(this.age>o.age) {
        return 1;
    }else {
        if(this.score<o.score) {
            return 1;
        }else if(this.score>o.score) {
            return -1;
        }else {
            return 0;
        }
    }
}

/**
 * hashCode方法和equals方法主要用来区分不同的对象，比如有两行内容如下：
 * zs  19  99
 * zs  19  80
 * 两者计算得到的hashCode()的值是不一样的，同时equals()方法的返回值也是false
 * 所以这两行对应的两个Student对象自然就不一样
 *key为Student类型时，其分组自然也不一样
 */
public int hashCode() {
    return this.name.hashCode()+this.age+this.score;
}
public boolean equals(Object obj) {
```

```
            if(!(obj instanceof Student)){
                return false;
            }
            Student s=(Student)obj;
                return this.name.equals(s.name)&&(this.age==s.age)&&(this.score==s.score);
        }

        /**
         * Reduce 中的 context.write(k2,NullWritable.get()) 方法是
         *把对象 k2(Student) 和 NullWritable.get() 作为键值对
         *默认通过 Tab 键作为分隔，输出到 part-r-0000* 文件中
         *这里 value 设置为空，所以只输出 k2，所以重写
         *toString() 的目的就是把 Student 对象按照下面的格式输出到文件中
         * 当然也可以不重写该方法
         *那么就需要在 Reduce 中的 context.write() 中自己定义格式
         *比如 context.write(k2.getName+"\t"+k2.getAge()+"\t"+k2.getScore())
         */
        public String toString() {
            return name+"\t"+age+"\t"+score;
        }
    }
}
```

把代码打成 jar 包执行。

```
shiyanlou:~/ $ hadoop jar CustomSort.jar org.lanqiao.bigdata.CustomSort /input /out
```

执行完毕后，查看输出结果。

```
shiyanlou:~/ $ hadoop fs -cat /out/part-r-00000
sq   19   88
zs   20   100
qb   20   96
zl   21   99
ls   21   95
ww   22   90
```

可以看到，程序成功在不改变列的顺序的前提下，先按照年龄升序排列，再按照成绩降序排列。

3．Shuffle 规约

下面我们举一个例子，来实现一个 Shuffle 规约的功能。

首先启动 HDFS 和 Yarn。

```
shiyanlou:~/ $ start-all.sh
……
```

准备数据文件 a.txt，分隔符为空格。

```
shiyanlou:~/ $ cat a.txt
i you his
you her
his it
```

然后把该文件上传到 HDFS 上的 /input 目录下。

```
shiyanlou:~/ $ hadoop fs -mkdir /input
shiyanlou:~/ $ hadoop fs -put a.txt /input/
```

该文件含有一些单词，现在使用 Shuffle 规约功能来统计单词数量。这相当于在 map 端先提前汇总一次，然后在 reduce 端再汇总一次，提高效率，减少 map 端发送给 reduce 端的数据量，减轻带宽压力。

使用 Eclipse 作为 IDE，保持开发环境和之前实验环境一样就可以了，详细代码如程序清单 5.4 所示。

程序清单 5.4

```java
package org.lanqiao.bigdata;

import java.net.URI;

import org.apache.hadoop.conf.Configuration;
import org.apache.hadoop.fs.FileSystem;
import org.apache.hadoop.fs.Path;
import org.apache.hadoop.io.LongWritable;
import org.apache.hadoop.io.Text;
import org.apache.hadoop.mapreduce.Job;
import org.apache.hadoop.mapreduce.Mapper;
import org.apache.hadoop.mapreduce.Reducer;
import org.apache.hadoop.mapreduce.lib.input.FileInputFormat;
import org.apache.hadoop.mapreduce.lib.output.FileOutputFormat;

public class CustomCombine {
    public static void main(String[] args) throws Exception {
        Configuration conf=new Configuration();
        // 获取集群访问路径
        FileSystem fileSystem=FileSystem.get(new URI("hdfs://localhost:9000/"),conf);
```

```java
            // 判断输出路径是否已经存在,如果存在则删除
            Path outPath=new Path(args[1]);
            if(fileSystem.exists(outPath)){
    fileSystem.delete(outPath,true);
            }
            // 生成job,并指定job的名称
            Job job=Job.getInstance(conf,"custom-combine");
            // 指定打成jar后的运行类
    job.setJarByClass(CustomCombine.class);
            // 指定mapper类
    job.setMapperClass(MyMapper.class);
            // 指定reducer类
    job.setReducerClass(MyReducer.class);
            // 指定mapper的输出类型
    job.setMapOutputKeyClass(Text.class);
    job.setMapOutputValueClass(LongWritable.class);
            // 指定reducer的输出类型
    job.setOutputKeyClass(Text.class);
    job.setOutputValueClass(LongWritable.class);

            // 局部规约
    job.setCombinerClass(MyCombiner.class);

    FileInputFormat.addInputPath(job,new Path(args[0]));
    FileOutputFormat.setOutputPath(job,new Path(args[1]));
    System.exit(job.waitForCompletion(true) ? 0:1);
        }

    static class MyMapper extends Mapper<LongWritable,Text,Text,LongWritable>{
            protected void map(LongWritable k1,Text v1,Context context) throws java.io.IOException,InterruptedException {
                final String[] splited=v1.toString().split("");
                for (String word:splited) {
    context.write(new Text(word),new LongWritable(1));
                }
            };
        }
```

```
        static class MyReducer extends Reducer<Text,LongWritable,Text,LongWritab
le>{
            protected void reduce(Text k2,java.lang.Iterable<LongWritable>
v2s,Context ctx) throws java.io.IOException ,InterruptedException {
                long times=0L;
                for (LongWritable count:v2s) {
                    times+=count.get();
                }
    ctx.write(k2,new LongWritable(times));
            };
        }


        static class MyCombiner extends Reducer<Text,LongWritable,Text,LongWritable>{
            protected void reduce(Text k2,java.lang.Iterable<LongWritable>
v2s,Context ctx) throws java.io.IOException ,InterruptedException {
                long times=0L;
                for (LongWritable count:v2s) {
                    times+=count.get();
                }
    ctx.write(k2,new LongWritable(times));
            };
        }
    }
```

把代码打成 jar 包执行。

```
shiyanlou:~/ $ hadoop jar CustomCombine.jar org.lanqiao.bigdata.CustomCombine /input /out
```

执行完毕后，查看输出结果。

```
shiyanlou:~/ $ hadoop fs -cat /out/part-r-00000
her 1
his 2
i   1
it  1
you 2
```

可以看到它与不使用 Combiner 统计的结果一样，都是正确的。

附录 5　Flume 安装部署和配置

数据采集我们使用 Flume。下面先来看一下 Flume 的安装部署和配置。

首先登录 Flume 的官方网站下载 apache-flume-1.9.0-bin.tar.gz，将其上传到 /opt 目录下，再解压缩并重命名。

```
shiyanlou:/opt/ $ tar -zxvf apache-flume-1.9.0-bin.tar.gz
shiyanlou:/opt/ $ mv apache-flume-1.9.0-bin flume
```

然后修改配置文件。

```
shiyanlou:/flume/ $ cd conf/
shiyanlou:/conf/ $ cp flume-env.sh.template flume-env.sh
shiyanlou:/conf/ $ vi flume-env.sh
……
export JAVA_HOME=/opt/jdk1.8
……
```

主要是配置 JAVA_HOME。

最后配置环境变量。

```
shiyanlou:/conf/ $ sudo vi ~/.zshrc
……
export FLUME_HOME=/opt/flume
export PATH=$FLUME_HOME/bin:$PATH
shiyanlou:/conf/ $ source ~/.zshrc
```

Flume 框架对 Hadoop 的依赖只是在 jar 包上，并不要求启动 Flume 时必须将 Hadoop 服务也启动。

下面我们来验证一下 Flume 的安装是否成功。

```
shiyanlou:/conf/ $ flume-ng version
Flume 1.9.0
Source code repository:https://git-wip-us.apache.org/repos/asf/flume.git
Revision:d4fcab4f501d41597bc616921329a4339f73585e
Compiled by fszabo on Mon Jan 05 10:45:25 CET 2021
From source with checksum 35db629a3bda49d23e9b3690c80737f9
```

从输出可知安装成功。

附录6 Hive 安装部署和配置

数据分析我们使用 Hive。下面先进行 Hive 的安装部署和配置。

首先下载 Hive，这里我们登录官方网站下载 apache-hive-2.3.4-bin.tar.gz，解压后重命名。

```
shiyanlou:~/ $ tar -zxvf apache-hive-2.3.4-bin.tar.gz
shiyanlou:~/ $ mv apache-hive-2.3.4-bin hive-2.3.4
```

进入 conf 目录，创建 hive-site.xml 文件，然后添加 MySQL 的连接信息，详细代码如程序清单 5.5 所示。

程序清单 5.5

```
shiyanlou:~/ $ cd hive-2.3.4/conf
shiyanlou:~/ $ vi hive-site.xml
shiyanlou:hive-2.3.4/ $ cd conf
shiyanlou:conf/ $ vi hive-site.xml
<configuration>
    # 配置 MySQL 的连接地址、连接数据库
    <property>
        <name>javax.jdo.option.ConnectionURL</name>
        <value>jdbc:mysql://localhost:3306/hivedb?createDatabaseIfNotExist=true</value>
    </property>
    # 配置 MySQL 的驱动类
    <property>
        <name>javax.jdo.option.ConnectionDriverName</name>
        <value>com.mysql.jdbc.Driver</value>
    </property>
    # 配置登录 MySQL 的用户名
    <property>
        <name>javax.jdo.option.ConnectionUserName</name>
        <value>root</value>
    </property>
    # 配置登录 MySQL 的密码
    <property>
        <name>javax.jdo.option.ConnectionPassword</name>
        <value>123456</value>
    </property>
</configuration>
```

把 MySQL 驱动包（mysql-connector-java-5.1.35-bin.jar）放置在 Hive 的根路径下的 lib 目录中。

```
shiyanlou:~/ $ cp mysql-connector-java-5.1.35-bin.jar hive-2.3.4/lib/
```

启动 MySQL 服务，初始化元数据库。

```
shiyanlou:~/ $ service mysql start
shiyanlou:~/ $ schematool -dbTypemysql -initSchema
```

在 MySQL 数据库中就多了一个 hivedb 的数据库，这里面有很多与 Hive 元数据相关的表。

启动 Hive 进行验证。由于 Hive 要使用 Hadoop 的 HDFS 和 Yarn，所以需要先把 Hadoop 集群启动。每个 Hive 会话都会启动一个 JVM 来处理，因此我们直接使用 hive 命令，它相当于 hive --service cli 命令。

```
shiyanlou:~/ $ hive
……
hive>show databases;
default
```

至此，Hive 安装结束，我们可以看到默认有一个名为 default 的数据库。

附录 7　Sqoop 安装部署和配置

导出分析结果到 MySQL，我们需要使用 Sqoop。下面先来看下 Sqoop 的安装部署和配置。

1．下载

首先登录官方网站下载 sqoop-1.4.7.bin__hadoop-2.6.0.tar.gz，然后把该文件上传到 /opt 目录下。

2．解压缩并重命名

```
shiyanlou:/opt/ $ tar -zxvfsqoop-1.4.7.bin__hadoop-2.6.0.tar.gz
shiyanlou:/opt/ $ mv sqoop-1.4.7.bin__hadoop-2.6.0 sqoop-1.4.7
```

3．修改配置文件

修改配置文件的详细代码如程序清单 5.6 所示。

程序清单 5.6

```
shiyanlou:/opt/ $ cd sqoop-1.4.7/conf
shiyanlou:/conf/ $ mv sqoop-env-template.sh sqoop-env.sh
shiyanlou:/conf/ $ vi sqoop-env.sh
#Set path to where bin/hadoop is available
export HADOOP_COMMON_HOME=/opt/hadoop-2.9.2

#Set path to where hadoop-*-core.jar is available
export HADOOP_MAPRED_HOME=/opt/hadoop-2.9.2

#set the path to where bin/hbase is available
#export HBASE_HOME=

#Set the path to where bin/hive is available
export HIVE_HOME=/opt/hive-2.3.4

#Set the path for where zookeper config dir is
export ZOOCFGDIR=/opt/zookeeper-3.4.5/conf
```

4．加载 MySQL 驱动

把 MySQL 的驱动复制到 sqoop-1.4.7/lib 目录下即可。

```
shiyanlou:/opt/ $ cp hive-2.3.4/lib/mysql-connector-java-5.1.35-bin.jar sqoop-1.4.7/lib/
```

5．配置环境变量

```
shiyanlou:/opt/ $ sudo vi ~/.zshrc
export SQOOP_HOME=/opt/sqoop-1.4.7
export PATH=$SQOOP_HOME/bin:$PATH
shiyanlou:/opt/ $ source ~/.zshrc
```

6．验证安装是否成功

```
shiyanlou:/opt/ $ sqoop version
Warning:/opt/sqoop-1.4.7/../hbase does not exist! HBase imports will fail.
Please set $HBASE_HOME to the root of your HBase installation.
Warning:/opt/sqoop-1.4.7/../hcatalog does not exist! HCatalog jobs will fail.
Please set $HCAT_HOME to the root of your HCatalog installation.
Warning:/opt/sqoop-1.4.7/../accumulo does not exist! Accumulo imports will fail.
Please set $ACCUMULO_HOME to the root of your Accumulo installation.
21/01/07 10:11:16 INFO sqoop.Sqoop:Running Sqoop version:1.4.7
Sqoop 1.4.7
```

```
git commit id 2328971411f57f0cb683dfb79d19d4d19d185dd8
Compiled by maugli on Thur Jan 07 10:11:17 STD 2021
```

可以看到安装成功，几个警告提示可以暂时不做处理。

附录 8　Hadoop 高可用集群环境安装部署和配置

集群部署节点的角色规划如下（3 节点）。

```
server1：namenode、resourcemanager、zkfc、nodemanager、datanode、QuorumPeerMain、
JournalNode
server2：namenode、resourcemanager、zkfc、nodemanager、datanode、QuorumPeerMain、
JournalNode
server3：datanode、nodemanager、QuorumPeerMain、JournalNode
```

具体的安装步骤如下。

1. 安装配置 Zookeeper 集群

（1）解压缩。

```
[hadoop@server1 ~]$ tar -zxvf zookeeper-3.4.5.tar.gz -C /home/hadoop/app/
```

（2）修改配置。

```
[hadoop@server1 conf]$ cd /home/hadoop/app/zookeeper-3.4.5/conf/
[hadoop@server1 conf]$ cp zoo_sample.cfgzoo.cfg
[hadoop@server1 conf]$ vim zoo.cfg
……
dataDir=/home/hadoop/app/zookeeper-3.4.5/tmp# 主要修改此处，添加后面 3 行配置

server.1=server1:2888:3888
server.2=server2:2888:3888
server.3=server3:2888:3888
```

（3）创建一个 tmp 文件夹，把数字 1 存入一个新文件 myid 中。

```
[hadoop@server1 conf]$ mkdir /home/hadoop/app/zookeeper-3.4.5/tmp
[hadoop@server1 conf]$ echo 1 > /home/hadoop/app/zookeeper-3.4.5/tmp/myid
```

（4）将配置好的 Zookeeper 拷贝到其他节点（server2、server3）上。

```
[hadoop@server1 conf]$ scp -r /home/hadoop/app/zookeeper-3.4.5/ server2:/home/hadoop/app/
[hadoop@server1 conf]$ scp -r /home/hadoop/app/zookeeper-3.4.5/ server3:/home/hadoop/app/
```

（5）修改 server2、server3 对应 myid 内容为 2 和 3。

```
[hadoop@server2 conf]$ echo 2> /home/hadoop/app/zookeeper-3.4.5/tmp/myid
[hadoop@server3 conf]$ echo 3> /home/hadoop/app/zookeeper-3.4.5/tmp/myid
```

2. 安装配置 Hadoop 集群

（1）解压缩。

```
[hadoop@server1 ~]$ tar -zxvf hadoop-2.6.0.tar.gz -C /home/hadoop/app/
```

（2）配置 HDFS。

① 将 Hadoop 添加到环境变量中。

```
[hadoop@server1 ~]$ vim /etc/profile
……
export JAVA_HOME=/home/hadoop/app/jdk1.8
export HADOOP_HOME=/home/hadoop/app/hadoop-2.6.0
export PATH=$PATH:$JAVA_HOME/bin:$HADOOP_HOME/bin
```

② 修改 hadoop-env.sh。

```
[hadoop@server1 ~]$ cd /home/hadoop/app/hadoop-2.6.0/etc/hadoop
[hadoop@server1 hadoop]$ vim hadoop-env.sh
export JAVA_HOME=/home/hadoop/app/jdk1.8
……
```

③ 修改 core-site.xml，详细代码如程序清单 5.7 所示。

程序清单 5.7

```xml
[hadoop@server1 hadoop]$ vim core-site.xml
<configuration>
    <!-- 指定 HDFS 的 nameservice -->
    <property>
        <name>fs.defaultFS</name>
        <value>hdfs://ns1</value>
    </property>
    <!-- 指定 Hadoop 临时目录 -->
    <property>
        <name>hadoop.tmp.dir</name>
        <value>/home/hadoop/app/ hadoop-2.6.0/tmp</value>
    </property>
    <!-- 指定 Zookeeper 地址 -->
    <property>
        <name>ha.zookeeper.quorum</name>
        <value>server1:2181,server2:2181,server3:2181</value>
    </property>
</configuration>
```

④ 修改 hdfs-site.xml，详细代码如程序清单 5.8 所示。

程序清单 5.8

```
[hadoop@server1 hadoop]$ vim hdfs -site.xml
<configuration>
    <!-- 指定 HDFS 副本数为 3 -->
    <property>
        <name>dfs.replication</name>
        <value>3</value>
    </property>
    <!-- 指定 HDFS 的 nameservice 为 ns1，需要和 core-site.xml 中的保持一致 -->
    <property>
        <name>dfs.nameservices</name>
        <value>ns1</value>
    </property>
    <!-- ns1 下面有两个 NameNode，分别是 nn1, nn2 -->
    <property>
        <name>dfs.ha.namenodes.ns1</name>
        <value>nn1,nn2</value>
    </property>
    <!-- nn1 的 RPC 通信地址 -->
    <property>
        <name>dfs.namenode.rpc-address.ns1.nn1</name>
        <value>server1:9000</value>
    </property>
    <!-- nn1 的 http 通信地址 -->
    <property>
        <name>dfs.namenode.http-address.ns1.nn1</name>
        <value>server1:50070</value>
    </property>
    <!-- nn2 的 RPC 通信地址 -->
    <property>
        <name>dfs.namenode.rpc-address.ns1.nn2</name>
        <value>server2:9000</value>
    </property>
    <!-- nn2 的 http 通信地址 -->
    <property>
        <name>dfs.namenode.http-address.ns1.nn2</name>
        <value>server2:50070</value>
    </property>
    <!-- 指定 NameNode 的 edits 元数据在 JournalNode 上的存放位置 -->
```

```xml
    <property>
            <name>dfs.namenode.shared.edits.dir</name>
            <value>qjournal://server1:8485;server2:8485;server3:8485/ns1</value>
    </property>
    <!-- 指定 JournalNode 在本地磁盘存放数据的位置 -->
    <property>
            <name>dfs.journalnode.edits.dir</name>
            <value>/home/hadoop/app/hadoop-2.6.0/journaldata</value>
    </property>
    <!-- 开启 NameNode 失败自动切换 -->
    <property>
            <name>dfs.ha.automatic-failover.enabled</name>
            <value>true</value>
    </property>
    <!-- 配置失败自动切换实现方式 -->
    <property>
            <name>dfs.client.failover.proxy.provider.ns1</name>
            <value>org.apache.hadoop.hdfs.server.namenode.ha.ConfiguredFailoverProxyProvider</value>
    </property>
    <!-- 配置隔离机制方法，多个机制用换行分割，即每个机制暂用一行 -->
    <property>
            <name>dfs.ha.fencing.methods</name>
            <value>sshfence</value>
    </property>
    <!-- 使用 sshfence 隔离机制时需要 SSH 免登录 -->
    <property>
            <name>dfs.ha.fencing.ssh.private-key-files</name>
            <value>/home/hadoop/.ssh/id_rsa</value>
    </property>
    <!-- 配置 sshfence 隔离机制超时时间 -->
    <property>
            <name>dfs.ha.fencing.ssh.connect-timeout</name>
            <value>30000</value>
    </property>
</configuration>
```

⑤ 修改 mapred-site.xml，详细代码如程序清单 5.9 所示。

程序清单 5.9

```
[hadoop@server1 hadoop]$ vim mapred-site.xml
<configuration>
```

```xml
<!-- 指定 MapReduce 框架为 Yarn 方式 -->
<property>
     <name>mapreduce.framework.name</name>
     <value>yarn</value>
</property>
</configuration>
```

⑥ 修改 yarn-site.xml，详细代码如程序清单 5.10 所示。

程序清单 5.10

```
[hadoop@server1 hadoop]$ vim yarn-site.xml
<configuration>
     <!-- 开启 ResourceManager 高可用 -->
     <property>
          <name>yarn.resourcemanager.ha.enabled</name>
          <value>true</value>
     </property>
     <!-- 指定 ResourceManager 的 cluster id -->
     <property>
          <name>yarn.resourcemanager.cluster-id</name>
          <value>yrc</value>
     </property>
     <!-- 指定 ResourceManager 的名字 -->
     <property>
          <name>yarn.resourcemanager.ha.rm-ids</name>
          <value>rm1,rm2</value>
     </property>
     <!-- 分别指定 ResourceManager 的地址 -->
     <property>
          <name>yarn.resourcemanager.hostname.rm1</name>
          <value>server1</value>
     </property>
     <property>
          <name>yarn.resourcemanager.hostname.rm2</name>
          <value>server2</value>
     </property>
     <!-- 指定 Zookeeper 集群地址 -->
     <property>
          <name>yarn.resourcemanager.zk-address</name>
          <value>server1:2181,server2:2181,server3:2181</value>
     </property>
     <property>
```

```
            <name>yarn.nodemanager.aux-services</name>
            <value>mapreduce_shuffle</value>
    </property>
</configuration>
```

⑦ 修改 slaves。

```
[hadoop@server1 hadoop]$ vim slaves
server1
server2
server3
```

（3）将配置好的 Hadoop 拷贝到其他节点。

```
[hadoop@server1 ~]$ scp -r /home/hadoop/app/hadoop-2.6.0/ hadoop@server2:/home/hadoop/app/
[hadoop@server1 ~]$ scp -r /home/hadoop/app/hadoop-2.6.0/ hadoop@server3:/home/hadoop/app/
```

（4）配置集群 SSH 免密码登录。

① 分别在 3 个节点 server1、server2、server3 上产生秘钥。

```
[hadoop@server1 ~]$ ssh-keygen -t rsa
[hadoop@server2 ~]$ ssh-keygen -t rsa
[hadoop@server3 ~]$ ssh-keygen -t rsa
```

② 将各自的公钥拷贝到其他节点上，包括自己。

```
[hadoop@server1 ~]$ ssh-copy-id server1
[hadoop@server1 ~]$ ssh-copy-id server2
[hadoop@server1 ~]$ ssh-copy-id server3
[hadoop@server2 ~]$ ssh-copy-id server1
[hadoop@server2 ~]$ ssh-copy-id server2
[hadoop@server2 ~]$ ssh-copy-id server3
[hadoop@server3 ~]$ ssh-copy-id server1
[hadoop@server3 ~]$ ssh-copy-id server2
[hadoop@server3 ~]$ ssh-copy-id server3
```

3．启动验证

（1）启动 ZooKeeper 集群。

分别在 server1、server2 和 server3 节点上启动 ZooKeeper 集群。

```
[hadoop@server1 ~]$ /home/hadoop/app/zookeeper-3.4.5/bin/zkServer.sh start
[hadoop@server2 ~]$ /home/hadoop/app/zookeeper-3.4.5/bin/zkServer.sh start
[hadoop@server3 ~]$ /home/hadoop/app/zookeeper-3.4.5/bin/zkServer.sh start
```

（2）启动 journalnode 进程。

分别在 server1、server2 和 server3 节点上启动 journalnode 进程。

```
[hadoop@server1 ~]$ /home/hadoop/app/hadoop-2.6.0/sbin/hadoop-daemon.sh start journalnode
[hadoop@server2 ~]$ /home/hadoop/app/hadoop-2.6.0/sbin/hadoop-daemon.sh start journalnode
[hadoop@server3 ~]$ /home/hadoop/app/hadoop-2.6.0/sbin/hadoop-daemon.sh start journalnode
```

（3）格式化 HDFS。

在 server1 节点上执行以下命令。

```
[hadoop@server1 ~]$ /home/hadoop/app/hadoop-2.6.0/bin/hdfsnamenode -format
```

格式化后会根据 core-site.xml 中的 hadoop.tmp.dir 配置生成文件，这里配置的是 /home/hadoop/app/ hadoop-2.6.0/tmp，然后将 /home/hadoop/app/ hadoop-2.6.0/tmp 拷贝到 server2 节点的 /home/hadoop/app/ hadoop-2.6.0/ 下即可。

```
[hadoop@server1 ~]$ scp -r /home/hadoop/app/hadoop-2.6.0/tmpserver2:/home/hadoop/app/hadoop-2.6.0/
```

如果不想复制也可以这样操作，先在 server1 节点上单独启动 namenode。

```
[hadoop@server1 ~]$ /home/hadoop/app/hadoop-2.6.0/sbin/hadoop-daemon.sh startnamenode
```

然后在 server2 节点上执行以下命令。

```
[hadoop@server2 ~]$ /home/hadoop/app/hadoop-2.6.0/bin/hdfsnamenode -bootstrapStandby
```

（4）格式化 ZKFC。

在 server1 节点上执行以下命令。

```
[hadoop@server1 ~]$ /home/hadoop/app/hadoop-2.6.0/bin/hdfszkfc-formatZK
```

（5）启动 HDFS。

在 server1 节点上执行以下命令。

```
[hadoop@server1 ~]$ /home/hadoop/app/hadoop-2.6.0/sbin/start-dfs.sh
```

这样 server1 上的 NameNode 默认就是 active 状态的，server2 上的 NameNode 默认就是 standby 状态的。

（6）启动 Yarn。

在 server2 节点上执行以下命令。

```
[hadoop@server2 ~]$ /home/hadoop/app/hadoop-2.6.0/sbin/start-yarn.sh
```

这样 server2 节点上的 ResourceManager 默认就是 active 状态，server1 节点上的 ResourceManager 默认就是 standby 状态。

启动后 server2 节点上有一个 ResourceManager 进程，而 server1 节点上没有 ResourceManager 进程，需要手动启动。

```
[hadoop@server1 ~]$ /home/hadoop/app/hadoop-2.6.0/sbin/yarn-daemon.sh start resourcemanager
```

到此，Hadoop HA 启动完毕，在 server1、server2、server3 节点上分别查看进程。

```
[hadoop@server1 ~]$ jps
6229 DFSZKFailoverController
1303 QuorumPeerMain
5879 DataNode
6616 Jps
6073 JournalNode
6443 NodeManager
5774 NameNode
2920 ResourceManager
[hadoop@server2 ~]$ jps
3812 DFSZKFailoverController
3575 NameNode
3927 NodeManager
4040 Jps
1291 QuorumPeerMain
3723 JournalNode
2750 ResourceManager
3646 DataNode
[hadoop@server3 ~]$ jps
1939 NodeManager
2857 Jps
4521 QuorumPeerMain
2349 JournalNode
3132 DataNode
```

可以看到 server1 和 server2 节点上各有 7 个进程，server3 节点上有 4 个进程，符合初始设定。

（7）验证 NameNode HA。

验证思路：首先查看哪个节点是 active 状态，哪个节点是 standby 状态。然后把 active 状态的节点的 NameNode 进程关闭，看 standby 状态的节点能否自动变成 active 状态，同时验证 HDFS 是否可用。

首先查看 server1 和 server2 节点的 NameNode 信息。

```
http://server1:50070
NameNode 'server1:9000' (active)
http://server2:50070
NameNode 'server2:9000' (standby)
```

可以明显地看到各自的状态。

然后向 HDFS 上传一个文件。

```
[hadoop@server1 ~]$ hadoop fs -mkdir/input
[hadoop@server1 ~]$ hadoop fs -put /a.txt/input
[hadoop@server1 ~]$ hadoop fs -ls /input
-rw-r--r--   3 hadoop supergroup       1926 2022-03-23 15:36  /input/a.txt
```

再关闭 active 状态的 NameNode 进程。

```
[hadoop@server1 ~]$ kill -9 5774
```

通过浏览器访问以下地址。

```
http://server1:50070
```

发现已经不能访问了。

再次通过浏览器访问以下地址。

```
http://server2:50070
NameNode 'server2:9000' (active)
```

这个时候 server2 节点上的 NameNode 进程的状态变成了 active 状态。

再次执行以下命令。

```
[hadoop@server1 ~]$ hadoop fs -ls /input
-rw-r--r--   3 hadoop supergroup       1926 2022-03-23 15:36  /input/a.txt
```

刚才上传的文件依然存在。

手动启动关闭的 NameNode 进程。

```
[hadoop@server1 ~]$ /home/hadoop/app/hadoop-2.6.0/sbin/hadoop-daemon.sh start namenode
```

通过浏览器访问以下地址。

```
http://server1:50070
NameNode 'server1:9000' (standby)
```

这时 server1 节点上的 NameNode 进程的状态变成了 standby 状态。

（8）验证 Yarn。

首先运行以下 Hadoop 提供的 demo 中的 WordCount 程序。

```
[hadoop@server1 ~]$ /home/hadoop/app/hadoop-2.6.0/bin/hadoop
jar /home/hadoop/app/hadoop-2.6.0/share/hadoop/mapreduce/hadoop-mapreduce-
examples-2.6.0.jar wordcount /input /out
```

然后快速关闭 active 状态的 ResourceManager 进程。

```
[hadoop@server2 ~]$ kill -9 2750
```

观察 server1 节点上的 ResourceManager 进程是否可以自动访问。结果 server2 节点上的 ResourceManager Web 客户端已经不能访问，server1 节点的 ResourceManager 进程自动变为 active 状态。观察发现 MapreduceJob 能正常运行，ResourceManager 进程的意外故障未影响其运行。

也可以查看状态。

```
[hadoop@server1 ~]$ yarn rmadmin-getServiceState rm1
active
[hadoop@server2 ~]$ yarn rmadmin-getServiceState rm2
standby
```

附录 9　Hadoop 集群节点动态管理

1．Hadoop 集群动态添加节点

已知目前 Hadoop 集群的节点为 server1（master）、server2（slave）和 server3（slave），现在希望在不关闭集群的情况下，动态地添加节点 server4（slave），该如何做呢？

（1）配置 server4 节点。

保证 server4 节点和集群节点差不多，比如都安装了 JDK、Hadoop，配置好了 IP 地址和主机名，server1 节点可以 SSH 免密登录 server4 节点等。

（2）修改集群中所有节点的 hosts 文件，添加 server4 节点的 IP 与主机名映射。

```
[hadoop@server1 ~]$ vim /etc/hosts
192.168.128.101          server1
192.168.128.102          server2
192.168.128.103          server3
192.168.128.104          server4
```

在 server2、server3、server4 节点上进行同样的操作。

（3）修改集群中所有节点的 slaves 文件，添加 server4 节点。

```
[hadoop@server1 ~]$ vim /home/hadoop/app/hadoop-2.6.0/etc/hadoop/slaves
server1
server2
server3
server4
```

在 server2、server3、server4 节点上进行同样的操作。

（4）在 server4 节点上启动 DataNode 和 NodeManager 进程。

```
[hadoop@server4 ~]$ /home/hadoop/app/hadoop-2.6.0/sbin/hadoop-daemon.sh start datanode
[hadoop@server4 ~]$ /home/hadoop/app/hadoop-2.6.0/sbin/yarn-daemon.sh start nodemanager
```

（5）还可以选择在 server1 节点上设置 HDFS 负载均衡。

```
[hadoop@server1 ~]$ /home/hadoop/app/hadoop-2.6.0/sbin/start-balancer.sh
```

（6）验证。

打开浏览器，输入地址"http://server1:50070"和"http://server1:8088"，可以查看集群的 DataNode 和 NodeManager 进程的节点是否有所增加。

2．Hadoop 集群动态删除节点

已知目前 Hadoop 集群的节点为 server1（master）、server2（slave）、server3（slave）和 server4（slave），现在希望在不关闭集群的情况下，动态地删除节点 server4（slave），该如何做呢？

（1）修改集群中所有节点的 hosts 文件，删除 server4 节点的 IP 与主机名映射。

```
[hadoop@server1 ~]$ vim /etc/hosts
192.168.128.101         server1
192.168.128.102         server2
192.168.128.103         server3
```

（2）修改集群中所有节点的 slaves 文件，删除 server4 节点。

```
[hadoop@server1 ~]$ vim /home/hadoop/app/hadoop-2.6.0/etc/hadoop/slaves
server1
server2
server3
```

（3）在 server4 节点上停止 DataNode 和 NodeManager 进程。

```
[hadoop@server4 ~]$ /home/hadoop/app/hadoop-2.6.0/sbin/hadoop-daemon.sh stopdatanode
[hadoop@server4 ~]$ /home/hadoop/app/hadoop-2.6.0/sbin/yarn-daemon.sh stopnodemanager
```

(4)还可以选择在 server1 节点上设置 HDFS 负载均衡。

```
[hadoop@server1 ~]$ /home/hadoop/app/hadoop-2.6.0/sbin/start-balancer.sh
```

(5)验证。

打开浏览器,输入地址"http://server1:50070"和"http://server1:8088",可以查看集群的 DataNode 和 NodeManager 进程的节点是否有所减少。

附录 10　Kafka 安装部署和配置

Kafka 使用 kafka_2.11-2.4.1.tgz 版本,其中 2.11 是 Scala 版本,2.4.1 才是 Kafka 版本。使用 3 个节点安装 Kafka 集群,现已经安装好了 Java、Hadoop 和 ZooKeeper,具体规划如表附 10.1 所示。

表附 10.1　Kafka 集群部署规划

主机名	IP 地址	运行进程	软硬件配置
master	192.168.128.131	NameNode SecondaryNameNode ResourceManager QuorumPeerMain Kafka	操作系统:CentOS 7.6.1810 Java:Oracle JDK 8u65 Hadoop:Hadoop 2.9.2 内存:1GB CPU:1 个 1 核 硬盘:20GB ZooKeeper:ZooKeeper 3.4.5 HBase:1.4.13 Kafka:2.11-2.4.1
slave1	192.168.128.132	DataNode NodeManager QuorumPeerMain Kafka	操作系统:CentOS 7.6.1810 Java:Oracle JDK 8u65 Hadoop:Hadoop 2.9.2 内存:1GB CPU:1 个 1 核 硬盘:20GB ZooKeeper:ZooKeeper 3.4.5 HBase:1.4.13 Kafka:2.11-2.4.1
slave2	192.168.128.133	DataNode NodeManager QuorumPeerMain Kafka	操作系统:CentOS 7.6.1810 Java:Oracle JDK 8u65 Hadoop:Hadoop 2.9.2 内存:1GB CPU:1 个 1 核 硬盘:20GB ZooKeeper:ZooKeeper 3.4.5 HBase:1.4.13 Kafka:2.11-2.4.1

1. 下载 Kafka

首先到 Kafka 的官方网站上下载 kafka_2.11-2.4.1.tgz，然后上传到 3 个节点的 /soft 目录下。注意：后续需要在 3 个节点上做同样的操作，这里以 master 节点为例来演示。

2. 解压缩并重命名

```
[hadoop@master soft]$ tar -zxvf kafka_2.11-2.4.1.tgz
[hadoop@master soft]$ mv kafka_2.11-2.4.1 kafka
```

3. 修改配置文件

进入其 config 目录，准备修改 server.properties 文件。

```
[hadoop@mastersoft]$ cd kafka/config/
```

由于 server.properties 文件内容较多，这里不在原文件上直接修改参数，而是先备份原文件，然后新建一个配置文件，详细代码如程序清单 5.11 所示。

程序清单 5.11

```
[hadoop@masterconfig]$ mv server.properties server.properties.bak
[hadoop@masterconfig]$ vi server.properties
# 当前机器在集群中的唯一标识，和 zookeeper 的 myid 性质一样
broker.id=0
# 当前 Kafka 对外提供服务的端口默认是 9092
port=9092
# 这个是 borker 进行网络处理的线程数
num.network.threads=3
# 这个是 borker 进行 I/O 处理的线程数
num.io.threads=8
# 消息存放的目录，这个目录可以配置为用逗号分隔的表达式，data 目录会自动创建
log.dirs=/soft/kafka/data
# 发送缓冲区 buffer 大小，数据不是一下子就发送的，会先存储到缓冲区
# 到达一定的大小后再发送，能提高性能
socket.send.buffer.bytes=102400
# Kafka 接收缓冲区大小，当数据到达一定大小后再序列化到磁盘
socket.receive.buffer.bytes=102400
# 这个参数是向 Kafka 请求消息或者向 Kafka 发送消息的请求的最大数
# 这个值不能超过 Java 的堆栈大小
socket.request.max.bytes=104857600
# 默认的分区数，一个 topic 默认 1 个分区
num.partitions=3
# 默认消息的最大持久化时间——168 小时（7 天）
```

```
log.retention.hours=168
# 消息保存的最大值为 5M
message.max.byte=5242880
#Kafka 保存消息的副本数，如果一个副本失效了，另一个还可以继续提供服务
default.replication.factor=2
# 取消息的最大字节数
replica.fetch.max.bytes=5242880
# 以下参数意味着，因为 Kafka 的消息是以追加的形式落地到文件的
# 当超过这个值的时候，Kafka 会新建一个文件
log.segment.bytes=1073741824
# 每隔 300000 毫秒检查一次上面配置的 log 失效时间是否是 log.retention.hours=168
# 到目录中查看是否有过期的消息，如果有，删除
log.retention.check.interval.ms=300000
# 是否启用 log 压缩。一般不用启用，启用可以提高性能
log.cleaner.enable=false
# 真正删除 topic，否则只是标记为删除
delete.topic.enable=true
# 设置 ZooKeeper 的连接端口
zookeeper.connect=192.168.128.131:2181,192.168.128.132:2181,192.168.128.133:2181
```

注意：slave1 和 slave2 节点上的参数 broker.id 依次设置为 1、2，master 节点上的参数 broker.id 设置为 0，其余设置都一样。

4．配置环境变量

```
[hadoop@master soft]$ sudo vi /etc/profile
export KAFKA_HOME=/soft/kafka
export PATH=$KAFKA_HOME/bin:$PATH
[hadoop@master soft]$ source /etc/profile
```

5．启动 Kafka 进程

首先启动 ZooKeeper 集群，然后在 3 个节点上以后台进程的方式启动 Kafka。

```
[hadoop@master soft]$ kafka-server-start.sh -daemon $KAFKA_HOME/config/server.properties
[hadoop@master soft]$ jps
1574 QuorumPeerMain
3304 Kafka
[hadoop@slave1 soft]$ jps
1520 QuorumPeerMain
2292 Kafka
```

```
[hadoop@slave2 soft]$ jps
1513 QuorumPeerMain
2282 Kafka
```

成功启动后可以看到 ZooKeeper 和 Kafka 的进程。

6．关闭 Kafka 进程

可以使用如下命令来关闭 Kafka 进程。

```
[hadoop@master soft]$ kafka-server-stop.sh
```

附录 11 Spark 安装部署和配置

Spark 包括 3 种不同类型的集群部署方式，包括 Standalone、Spark on Mesos 和 Spark on Yarn。

这里我们采用 Standalone 模式，Spark 的版本为 spark-2.3.4-bin-hadoop2.7.tgz，其内置的 Scala 版本为 2.11.8。计划在 3 个节点上安装 Spark 集群，已经安装好了 Java、Hadoop 和 ZooKeeper，具体规划如表附 11.1 所示。

表附 11.1 Spark 集群部署规划

主机名	IP 地址	运行进程	软硬件配置
master	192.168.128.131	NameNode SecondaryNameNode ResourceManager QuorumPeerMain Master（Spark 主节点）	操作系统：CentOS 7.6.1810 Java：Oracle JDK 8u65 Hadoop：Hadoop 2.9.2 内存：1GB CPU：1 个 1 核 硬盘：20GB ZooKeeper：ZooKeeper 3.4.5 HBase：1.4.13 Spark：2.3.4
slave1	192.168.128.132	DataNode NodeManager QuorumPeerMain Worker（Spark 从节点）	

续表

主机名	IP 地址	运行进程	软硬件配置
slave2	192.168.128.133	DataNode NodeManager QuorumPeerMain Worker（Spark 从节点）	操作系统：CentOS 7.6.1810 Java：Oracle JDK 8u65 Hadoop：Hadoop 2.9.2 内存：1GB CPU：1个1核 硬盘：20GB ZooKeeper：ZooKeeper 3.4.5 HBase：1.4.13 Spark：2.3.4

1．下载

在 Spark 的官网上下载 spark-2.3.4-bin-hadoop2.7.tgz 安装包，上传到 3 个节点的 /soft 目录下。注意：后续需要在 3 个节点上做同样的操作，这里以 master 节点为例来演示。

2．解压缩并重命名

```
[hadoop@master soft]$ tar -zxvf spark-2.3.4-bin-hadoop2.7.tgz
[hadoop@master soft]$ mv spark-2.3.4-bin-hadoop2.7 spark-2.3.4
```

3．修改配置文件

进入其 config 目录，修改 spark-env.sh.template 文件。

```
[hadoop@master conf]$ mv spark-env.sh.template spark-env.sh
[hadoop@master conf]$ vi spark-env.sh
……
export JAVA_HOME=/soft/jdk1.8
export SPARK_MASTER_IP=192.168.128.131
export SPARK_MASTER_PORT=7077
export SPARK_WORKER_CORES=2
export SPARK_WORKER_MEMORY=1500m
```

其中 JAVA_HOME 表示 Java 的主目录。

SPARK_MASTER_IP 表示 Spark 集群的主节点 master 的 IP 地址。

SPARK_MASTER_PORT 表示 Spark 集群的主节点 master 的默认端口为 7077。

SPARK_WORKER_CORES 表示作业可用的 CPU 内核数量。如果不设置，就是默认的虚拟机的 CPU 内核数。

SPARK_WORKER_MEMORY 表示作业可使用的内存容量。如果不设置，就是默认的 1GB。注意：如果使用 GB 为单位，则不能使用小数，如 1.5GB 是错误的，可改成以

MB 为单位的这种写法，即 1500MB。

修改 slaves.template 文件。

```
[hadoop@master conf]$ mv slaves.template slaves
[hadoop@master conf]$ vi slaves
……
192.168.128.132
192.168.128.133
```

在 slaves 文件中添加从节点（worker 节点）的 IP 地址，一行一个。

4．配置环境变量

```
[hadoop@master conf]$ sudo vi /etc/profile
……
export SPARK_HOME=/soft/spark-2.3.4
export PATH=$SPARK_HOME/bin:$PATH
[hadoop@master conf]$ source /etc/profile
```

5．启动验证

首先启动 Hadoop，至少要启动 HDFS，Yarn 可以不启动。

然后在 master 节点上使用 start-all.sh 命令启动 Spark 集群。

```
[hadoop@master soft]$ spark-2.3.4/sbin/start-all.sh
```

注意：这里之所以带上命令的路径，一是因为我们只配置了 bin 目录的环境变量，没有配置 sbin 目录的环境变量；二是因为此命令和 Hadoop 的启动命令 start-all.sh 功能一样。

启动后，查看进程，结果如下。

```
[hadoop@master soft]$ jps
2646 NameNode
2858 SecondaryNameNode
3310 Master
3007 ResourceManager
[hadoop@slave1 soft]$ jps
2530 NodeManager
2681 Worker
2458 DataNode
[hadoop@slave2 soft]$ jps
2448 DataNode
2614 Worker
2520 NodeManager
```

可以看到在主节点 master 上有 Master 进程，在从节点 slave1 和 slave2 上有 Worker 进程。我们还可以访问 master 节点的 Web 界面：http://192.168.128.131:8080/。

6. 关闭集群

```
[hadoop@master soft]$ spark-2.3.4/sbin/stop-all.sh
```

使用 stop-all.sh 命令关闭 Spark 集群。

至此，Standalone 形式的 Spark 集群已经安装部署成功。

附录 12　Spark RDD 算子

算子即高阶函数。Spark RDD 算子有两种类型：Transformation（转换）和 Action（动作）。所有的转换都是延迟加载的，也就是说，它们并不会直接计算结果，只是记住这些数据集上的转换。只有当发生一个动作时，这些转换才会真正地运行。这种延迟设计让 Spark 集群的运行效率更高。

下面我们就来介绍一些常用的转换和动作。为了方便讲解，我们以 Yarn 的 Spark Shell 的方式来学习。

进入客户端 Shell。

```
[hadoop@master ~]$ spark-shell --master yarn --deploy-mode client --executor-memory 1g
Spark context Web UI available at http://master:4040
……
scala>
```

可以看到此模式下 Web UI 界面的端口为 4040。

1. 创建 RDD

（1）parallelize() 方法。

在 Shell 下，可以使用 Scala 的集合作为数据集，使用 parallelize() 方法来创建 RDD。

```
scala>val rdd1=sc.parallelize(Array(1,2,3,4,5,6,7,8))
```

（2）textFile() 方法。

还可以用外部存储系统的数据集，包括本地的文件系统，以及所有 Hadoop 支持的数据集，如 HDFS、Cassandra、HBase 等作为数据集，使用 textFile() 方法来创建 RDD。

```
scala>val rdd2=sc.textFile("hdfs://192.168.128.131:9000/words.txt")
```

2．动作

前文我们学习了 2 种创建 RDD 的方法，但是不知道创建结果如何，下面我们来学习动作，它能使我们看见 RDD 的创建结果。

（1）collect( ) 方法。

collect( ) 方法将以数组的形式返回 RDD 的所有元素。

```
scala>rdd1.collect()
res0:Array[Int]=Array(1,2,3,4,5,6,7,8)
```

（2）count( ) 方法。

count( ) 方法将返回 RDD 的元素个数。

```
scala>rdd1.count()
res1:Long=8
```

（3）first( ) 方法。

first( ) 方法将返回 RDD 的第一个元素。

```
scala> rdd1.first()
res2:Int=1
```

（4）take(n) 方法。

take(n) 方法将以数组的形式返回 RDD 的前 n 个元素。

```
scala> rdd1.take(3)
res3:Array[Int]=Array(1,2,3)
```

（5）foreach(func) 方法。

foreach(func) 方法将对 RDD 中的每一个元素调用 func 函数计算，没有返回值。

```
scala>rdd1.foreach(_+1)
```

这里把 RDD 中每一个元素加 1，计算完毕之后，是看不到结果的，计算结果会在 Web UI 中的对应任务的 stdout 文件中输出。输出结果为 2 个文件，具体内容分别如下。

```
2
3
4
5
```

```
6
7
8
9
```

为什么有2个文件呢？这是因为如果没有设置RDD的分区数，那么其默认为CPU内核数，而笔者的计算机的CPU为2核。如果我们想要设置这个分区数，可以在创建时给parallelize( )方法添加第二个参数，如sc.parallelize(Array(1,2,3,4,5,6,7,8),3)，表示设置为3个分区。

（6）countByKey( ) 方法。

countByKey( ) 方法将针对 (K,V) 类型的 RDD，返回一个 (K,Int) 类型的 Map，表示每一个 key 对应的元素个数。

```
scala>sc.parallelize(List(("a",1),("b",2),("a",1))).countByKey()
res4:scala.collection.Map[String,Long]=Map(b -> 1,a -> 2)
```

（7）saveAsTextFile(path) 方法。

将 RDD 的元素以文本文件的形式保存到 HDFS 文件系统或其他支持的文件系统，对于每个元素，Spark 将会调用 toString 方法，将它转换为文件中的文本。

```
scala>rdd1.saveAsTextFile("hdfs://localhost: 9000/a")
[hadoop@master soft]$ hadoop fs -ls /a
-rw-r--r--   2 hadoop supergroup          0 2020-08-26 17:54/a/_SUCCESS
-rw-r--r--   2 hadoop supergroup          8 2020-08-26 17:54/a/part-00000
-rw-r--r--   2 hadoop supergroup          8 2020-08-26 17:54/a/part-00001
[hadoop@master soft]$ hadoop fs -cat /a/part-00000
2
3
4
5
[hadoop@master soft]$ hadoop fs -cat /a/part-00001
6
7
8
9
```

3．转换

下面介绍 RDD 的转换操作。

（1）map(func) 方法。

map(func) 方法将返回一个新的 RDD，该 RDD 由每一个输入元素经过 func 函数转换后组成。

```
scala>rdd1.map(_*2)
res5:org.apache.spark.rdd.RDD[Int]=MapPartitionsRDD[5] at map at <console>:26
```

可以看到转换并没有做任何计算，只是标记了它的计算和类型而已。如果要看结果，可以在它的后面调用一次动作。

```
scala>rdd1.map(_*2).collect()
res6:Array[Int]=Array(2,4,6,8,10,12,14,16)
```

（2）filter(func) 方法。

filter(func) 方法将返回一个新的 RDD，该 RDD 由经过 func 函数计算后返回值为 true 的输入元素组成。

```
scala>rdd1.filter(_>5).collect()
res7:Array[Int]=Array(6,7,8)
```

（3）flatMap(func) 方法。

如果把每一个输入元素映射为多个输出元素，则 map(func) 方法返回的结果就相当于一个二维数组。

```
scala>sc.parallelize(Array("hmm love lilei","tiger is big","huhu is big")).map(_.split(" ")).collect()
res8:Array[Array[String]]=Array(Array(hmm,love,lilei),Array(tiger,is,big),Array(huhu,is,big))
```

而 flatMap(func) 方法则会在此基础上把二维数组合并为一个一维数组。

```
scala>sc.parallelize(Array("hmm love lilei","tiger is big","huhu is big")).flatMap(_.split(" ")).collect()
res9:Array[String]=Array(hmm,love,lilei,tiger,is,big,huhu,is,big)
```

（4）mapPartitions(func) 方法。

在 map(func) 方法中，func 是作用在每个元素上的；在 mapPartitions(func) 方法中，func 是作用在每个分区上的。换句话说，map(func) 每次处理一条数据，而 mapPartitions(func) 每次处理一个分区的数据。这个分区的数据处理完后，原 RDD 中分区的数据才能释放，但这可能导致 OOM（内存溢出）。当内存空间较大时，建议使用 mapPartition(func) 方法，以提高处理效率。

```
scala>deffunc1(ite:Iterator[Int]):Iterator[Int]={for(it <-ite) yield it*it}
func1:(ite:Iterator[Int])Iterator[Int]
scala>rdd1.mapPartitions (func1).collect()
res10:Array[Int]=Array(1,4,9,16,25,36,49,64)
```

（5）mapPartitionsWithIndex(func) 方法。

mapPartitionsWithIndex(func) 方法的作用同 mapPartitions(func) 方法，只不过 func 多了一个分区索引的参数，可以更加细致地控制每个分区。

首先使用":paste"命令进入多行代码编写模式。

```
scala>:paste
// Entering paste mode (ctrl-D to finish)
import scala.collection.mutable.ListBuffer
def func2(index:Int,ite:Iterator[Int]):Iterator[Int]={
    var info=new ListBuffer[Int]()  // 收集每个分区的索引和数据
    var r=new ListBuffer[Int]()
    info += index
    for(it <- ite) {info += it; r += it*it;}
    println(info)
    r.toIterator
}
```

然后使用"Ctrl+D"组合键退出该模式。

```
scala>rdd1.mapPartitionsWithIndex(func2).collect()
res11:Array[Int]=Array(1,4,9,16,25,36,49,64)
```

这里显示的结果是计算后的返回值,而程序中 println(info) 的打印输出会在 Web UI 上对应任务的 stdout 中输出。输出结果有 2 个文件,具体内容分别如下。

```
ListBuffer(0,1,2,3,4)
```

```
ListBuffer(1,5,6,7,8)
```

(6) union(RDD) 方法。

union(RDD) 方法是对源 RDD 和参数 RDD 求并集后返回一个新的 RDD。

```
scala>var rdd1=sc.parallelize(1 to 10)
scala>var rdd2=sc.parallelize(5 to 15)
scala>rdd1.union(rdd2).collect()
res12:Array[Int]=Array(1,2,3,4,5,6,7,8,9,10,5,6,7,8,9,10,11,12, 13,14,15)
```

(7) intersection(RDD) 方法。

intersection(RDD) 方法是对源 RDD 和参数 RDD 求交集后返回一个新的 RDD。

```
scala>var rdd1=sc.parallelize(1 to 10)
scala>var rdd2=sc.parallelize(5 to 15)
scala>rdd1.intersection(rdd2).collect()
res13:Array[Int]=Array(6,8,10,7,9,5)
```

(8) distinct() 方法。

distinct() 方法是对源 RDD 进行去重后返回一个新的 RDD。

```
scala>var rdd1=sc.parallelize(1 to 10)
scala>var rdd2=sc.parallelize(5 to 15)
scala>rdd1.union(rdd2).distinct().collect()
res14:Array[Int]=Array(4,8,12,13,1,9,5,14,6,10,2,15,11,3,7)
```

（9）groupByKey( )方法。

groupByKey( )方法是在一个(K,V)类型的 RDD 上调用，返回一个依据 key 分组的(K,Iterator[V])类型的 RDD。

```
scala>sc.parallelize(List(("a",1),("b",1),("a",2),("c",3),("b",4))).groupByKey().collect()
res15:Array[(String,Iterable[Int])]=Array((c,CompactBuffer(3)),(a,CompactBuffer(1,2)),(b,CompactBuffer(1,4)))
```

（10）reduceByKey(func)方法。

reduceByKey(func)方法是在一个(K,V)类型的 RDD 上调用，返回一个(K,V)类型的 RDD。先依据 key 分组，然后使用指定的 func 函数对同一个组的元素进行计算。

```
scala>sc.parallelize(List(("a",1),("b",1),("a",2),("c",3),("b",4))).reduceByKey(_+_).collect()
res16:Array[(String,Int)]=Array((b,5),(a,3),(c,3))
```

（11）aggregate(initValue)(func1,func2)方法。

aggregate(initValue)(func1,func2)方法是一个聚合函数，接受多个输入，并按照一定的规则运算后输出一个结果值。该函数与分区有密切关系。

下面一起来看看如何查看 RDD 的分区数及分区情况。

① 查看 RDD 的分区数。

```
scala> rdd1.partitions.size
res17:Int=2
```

或者

```
scala> rdd1.getNumPartitions()
res18:Int=2
```

② 查看分区数据。

```
scala>rdd1.glom().collect()
res19:Array[Array[Int]]=Array(Array(1,2,3,4,5),Array(6,7,8,9,10))
```

可以看到 rdd1 分成 2 个分区，一个包含 1、2、3、4、5，另一个包含 6、7、8、9、10。

下面我们来看一个具体的例子：

```
scala>rdd1.aggregate(0)((x,y)=>math.max(x,y),(x,y)=>x+y)
res20:Int=15
```

该聚合函数有个初始值，包含 func1 和 func2 两个函数。其计算规则是，按照分区分别"滑动计算" func1 和 func2 函数。计算过程如下。

分区 1 中：初始值（0）与该分区中的第 1 个元素（1）按照 func1 函数规则比较大小，保留大数，即 1。然后用保留结果（1）继续和该分区中的第 2 个元素（2）按照 func1 函数规则比较大小，保留大数，即 2……就这样"滑动计算"，直到最后一个元素为止，结果为 5。

分区 2 中：与分区 1 同时计算，算法一样，初始值（0）与该分区中的第 1 个元素（6）按照 func1 函数规则比较大小，保留大数（6），然后用保留结果（6）继续和该分区中的第 2 个元素（7）按照 func1 函数规则比较大小，保留大数，即 7……就这样"滑动计算"，直到最后一个元素为止，结果为 10。

第 1 个函数 func1 的功能是在每个分区内遍历每个元素，将每个元素与初始值进行聚合。具体的聚合方式，我们可以自定义，不过有一点需要注意，聚合的时候依然要基于初始值来进行计算。然后按照 func2 函数规则来"滑动计算"最终结果：0+5+10=15。

第 2 个函数 func2 的功能是对每个分区聚合之后的结果进行再次合并，即分区之间的合并。同样，在合并的开始，也要基于初始值进行合并。

aggregate 函数是 Spark 中的一个高性能的算子，它实现了先进行分区内的聚合，然后对每个分区的聚合结果再次进行聚合的操作，这样在数据量较大的情况下，可大大减少数据在各个节点之间不必要的网络 IO（输入和输出），提升聚合性能。相比于 groupByKey 函数，在特定情况下，使用 aggregate 函数，聚合计算性能提升 10 倍不止。不过在使用的过程中一定要对该函数的每个参数的含义了如指掌，这样运用起来才能得心应手。

再看一个例子：

```
scala>val x=sc.parallelize(List("ab","abc","abcd","ab"))
scala>x.aggregate("")((x,y)=>math.max(x.length(),y.length()).toString(),(x,y)=>x+y)
```

同学们可以试试看，结果为 23 或者 32。

此外，还有一个与 aggregate 函数类似的 aggregateByKey 函数，二者用法差不多，只不过 aggregateByKey 函数只适用于 (K,V) 类型的数据，分区内 key 相同的一组才会聚合计算。

```
scala>val y=sc.parallelize(List(("a",1),("b",1),("a",2),("c",3),("a",4),("b",3),("c",1),("b",2)))
scala>y.aggregateByKey(2)(math.max(_ ,_), _ + _).collect()
res21:Array[(String,Int)]=Array((b,5),(a,6),(c,5))
```

同学们可以试试看。

（12）sortByKey(boolean) 方法。

sortByKey(boolean) 方法是在一个 (K,V) 类型的 RDD 上调用，K 必须实现 Ordered 接口，然后返回一个按照 key 进行排序的 (K,V) 类型的 RDD。参数 true 表示升序，参数 false 表示降序。

```
scala>val y=sc.parallelize(List(("a",1),("b",1),("a",2),("c",3),("a",4),("b",
3),("c",1),("b",2)))
  scala>y.sortByKey(true).collect()
  res22:Array[(String,Int)]=Array((a,4),(a,1),(a,2),(b,3),(b,2),(b,1),(c,3),(c,1))
  scala>y.sortByKey(false).collect()
  res23:Array[(String,Int)]=Array((c,3),(c,1),(b,3),(b,2),(b,1),(a,4),(a,1),(a,2))
```

(13) sortBy(func,boolean) 方法。

与 sortByKey(boolean) 方法类似,但其更加灵活,排序函数 func 需要自定义。

```
scala>sc.parallelize(1 to 10).sortBy(x=>x,false).collect()
res24:Array[Int]=Array(10,9,8,7,6,5,4,3,2,1)
scala>sc.parallelize(1 to 10).sortBy(x=>x+"",true).collect()
res25:Array[Int]=Array(1,10,2,3,4,5,6,7,8,9)
```

复合排序,即先按照学生的成绩降序排列,再按照年龄升序排列。

```
scala>var student=sc.parallelize(List(("zs",19,100),("ls",21,99),("ww",20,100),
("zl",19,99),("sq",21,100),("ldh",28,100),("hmm",20,99),("gfc",210,100)))
  scala>:paste
  // Entering paste mode (ctrl-D to finish)
  class Student(var name:String,var age:Int,var score:Int) extends
Ordered[Student] with Serializable {
    override def compare(that:Student):Int={
      if (this.score!=that.score) {
  that.score - this.score
      } else {
  this.age - that.age
      }
    }
  }
```

然后使用"Ctrl+D"组合键退出该模式。

```
scala>student.sortBy(x=>new Student(x._1,x._2,x._3)).collect
  res26:Array[(String,Int,Int)]=Array((zs,19,100),(ww,20,100),(sq,21,100),(ldh,2
8,100),(gfc,210,100),(zl,19,99),(hmm,20,99),(ls,21,99))
```

(14) join(RDD) 方法。

join(RDD) 方法是在类型为 (K,V) 和 (K,W) 的 RDD 上调用,返回一个 (K,(V,W)) 类型的 RDD。其中 K 为相同的 key,这个 key 对应的 V 和 W 排列组合,从而形成新的 RDD。

```
scala>var r1=sc.parallelize(List(("a",1),("b",1),("a",2),("c",3),("b",4)))
  scala>var r2=sc.parallelize(List(("a",10),("b",5),("c",3),("b",4)))
  scala>r1.join(r2).collect()
  res27:Array[(String,(Int,Int))]=Array((b,(4,4)),(b,(4,5)),(b,(1,4)),(b,(1,5)),
(a,(1,10)),(a,(2,10)),(c,(3,3)))
```

（15）coalesce(n,boolean) 与 repartition(n) 方法。

两个方法都是对数据重新分区。

repartition(n) 方法可以增加分区数，也可以减少分区数，其底部调用的是 coalesce(n,true)，其中 true 表示使用 shuffle。换句话说，就是无论分区数是增加还是减少，都会执行 shuffle 操作。所以如果是减少分区数，最好使用 coalesce(n,false) 方法。

```
scala>val rdd1=sc.parallelize(1 to 10,3)
scala>rdd1.glom().collect()
res28:Array[Array[Int]]=Array(Array(1,2,3),Array(4,5,6),Array(7,8,9,10))
scala>val rdd2=rdd1.coalesce(2,false)
scala>rdd2.glom().collect()
res29:Array[Array[Int]]=Array(Array(1,2,3),Array(4,5,6,7,8,9,10))
scala>val rdd2=rdd1.coalesce(2,true)
scala>rdd2.glom().collect()
res30:Array[Array[Int]]=Array(Array(5,1,3,7,9),Array(2,8,10,4,6))
```

可以看到，对于 coalesce 方法，不使用 shuffle 时，减少分区是直接合并部分分区；使用 shuffle 时，减少分区是打乱数据再分区。

```
scala>val rdd1=sc.parallelize(1 to 10,2)
scala>rdd1.glom().collect()
res31:Array[Array[Int]]=Array(Array(1,2,3,4,5),Array(6,7,8,9,10))
scala>val rdd2=rdd1.coalesce(3,false)
scala>rdd2.glom().collect()
res32:Array[Array[Int]]=Array(Array(1,2,3,4,5),Array(6,7,8,9,10))
scala>val rdd2=rdd1.coalesce(3,true)
scala>rdd2.glom().collect()
res33:Array[Array[Int]]=Array(Array(3,7,10),Array(8,1,4),Array(2,5,6,9))
```

可以看到，对于 coalesce 方法，不使用 shuffle 时，是不能增加分区的；使用 shuffle 时，增加分区是打乱数据再分区。

对于 repartition(n) 方法，我们就不再举例了，不管是增加分区，还是减少分区，都是调用使用 shuffle 的 coalesce 方法来完成的。

（16）partitionBy(Partitioner)。

partitionBy(Partitioner) 方法是对 (K,V) 类型的 RDD 进行自定义分区，参数需要继承 Partitioner。

首先定义分区函数。

```
scala>:paste
// Entering paste mode (ctrl-D to finish)
import org.apache.spark.Partitioner
class MyPartitoner(num:Int) extends Partitioner {
//var i:Int=-1
  override def getPartition(key:Any):Int={
    //方案一: hash 取模
val intKey=key.asInstanceOf[String].hashCode() //hash 取模
    var index=if(intKey>=0) {
intKey % num
    }else{
      (intKey+Int.MaxValue)%num
}
    index

    //方案二: 轮询
    //i+=1
    //var x=i%num
    //x
  }
}
```

然后使用"Ctrl+D"组合键退出该模式。

```
scala>val rdd1=sc.parallelize(List(("a",1),("b",1),("c",1),("d",1),("e",1),("f",1),("g",1),("h",1),("i",1),("j",1)))
  scala>rdd1.glom().collect()
  res34:Array[Array[(String,Int)]]=Array(Array((a,1),(b,1),(c,1),(d,1),(e,1)),Array((f,1),(g,1),(h,1),(i,1),(j,1)))
  scala>val rdd2=rdd1.partitionBy(new MyPartitoner(4))
  res35:Array[Array[(String,Int)]]=Array(Array((h,1),(d,1)),Array((a,1),(e,1),(i,1)),Array((f,1),(j,1),(b,1)),Array((c,1),(g,1)))
```

(17) foldByKey(initValue)(func) 方法。

foldByKey(initValue)(func) 方法是对 (K,V) 类型的 RDD 在分区内依据 key 分组后,按照 func 方法"滑动计算"。

```
scala>sc.parallelize(List(("a",1),("b",2),("a",3),("a",2),("b",1),("b",2))).foldByKey(10)(_+_).collect()
  res36:Array[(String,Int)]=Array((b,25),(a,26))
```

分区 1 内:a 的值为 10+1+3=14,b 的值为 10+2=12。

分区 2 内:a 的值为 10+2=12,b 的值为 10+1+2=13。

然后分区合并:a 的值为 14+12=26,b 的值为 12+13=25。

至此，我们挑选了一些非常重要又常用的 Spark 算子来学习，希望大家在课后多加练习。

附录 13　通过 WordCount 熟悉 Spark RDD

下面以 WordCount 为例，介绍其在 Spark 中的实现过程。

首先上传一个 words.txt 文件到 HDFS 中。

```
[hadoop@master soft]$ vi words.txt
hello world
hello hello world
kof
[hadoop@master soft]$ hadoop fs -put words.txt /
```

然后在 Spark Shell 中用 Scala 语言编写 Spark 程序，具体如下。

```
scala>sc.textFile("hdfs://localhost:9000/words.txt").flatMap(_.split(" ")).map((_,1)).reduceByKey(_+_).saveAsTextFile("hdfs://localhost:9000/out")
```

运行程序后，在 HDFS 上查看输出结果。

```
[hadoop@master soft]$ hadoop fs -cat /out/part*
(hello,3)
(kof,1)
(world,2)
```

上述程序中，sc 是 SparkContext 对象，该对象是提交 Spark 程序的入口。

textFile(hdfs://localhost:9000/words.txt) 是从 HDFS 中读取数据。

flatMap(_.split(" ")) 先使用空格，将每一行数据分隔成由多个单词构成的数组，然后把多行的数组合并成一个数组。

map((_,1)) 是将数组中每个单词和 1 构成元组映射。

reduceByKey(_+_) 是按照元组的 key 进行 reduce 操作，key 相同的元组的 value 将被累加起来。

saveAsTextFile("hdfs://localhost:9000/out") 将结果写入 HDFS 中的 out 目录。

可以看到使用 Spark 的原生语言 Scala 来编写 WordCount 非常简洁，采用链式方式，使用几个高阶函数（也叫算子）就可以实现了。如果使用 Java 语言来编写，代码量则是使用 Scala 语言的好几倍。每一个高阶函数要么使用匿名内部类，要么使用 Lambda 表达式，而这些实际上都属于函数式编程，所以使用 Scala 这种混合了函数式编程与面向对象编程的语言来编写代码会非常简洁。

附录 14　Flink 安装部署和配置

Flink 有 3 种不同类型的安装部署方式，即 Local、Standalone、Yarn 模式。

这里我们采用 Standalone 模式，Flink 的版本为 flink-1.10.2-bin-scala_2.11.tgz。计划在 3 个节点上安装 Flink 集群，已经安装好了 Java、Hadoop 和 ZooKeeper，具体规划如表附 14.1 所示。

表附 14.1　Flink 集群部署规划

主机名	IP 地址	运行进程	软硬件配置
master	192.168.128.131	NameNode SecondaryNameNode ResourceManager QuorumPeerMain Master（Flink 主节点）	操作系统：CentOS 7.6.1810 Java：Oracle JDK 8u65 Hadoop：Hadoop 2.9.2 内存：1GB CPU：1 个 1 核 硬盘：20GB ZooKeeper：ZooKeeper 3.4.5 Flink：1.10.2
slave1	192.168.128.132	DataNode NodeManager QuorumPeerMain Worker（Flink 从节点）	
slave2	192.168.128.133	DataNode NodeManager QuorumPeerMain Worker（Flink 从节点）	

1．下载

在 Flink 的官网上下载 flink-1.10.2-bin-scala_2.11.tgz 安装包，上传到 3 个节点的 /soft 目录下。注意：后续需要在 3 个节点上做同样的操作，这里以 master 节点为例来演示。

2．解压缩

```
[hadoop@master soft]$ tar -zxvf flink-1.10.2-bin-scala_2.11.tgz
```

3．修改配置文件

进入其 conf 目录。

```
[hadoop@master soft]$ cd flink-1.10.2/conf
```

修改 masters 文件，配置主节点。

```
[hadoop@masterconf]$ vi masters
master:8081
```

修改 slaves 文件，配置从节点。

```
[hadoop@masterconf]$ vi slaves
slave1
slave2
```

修改 flink-conf.yaml 文件，配置相关属性。

```
[hadoop@masterconf]$ vi flink-conf.yaml
jobmanager.rpc.address:master
taskmanager.numberOfTaskSlots:2
web.submit.enable:true
```

4．拷贝安装包到各节点

```
[hadoop@mastersoft]$ scp -r flink-1.10.2/ hadoop@slave1:/soft/
[hadoop@mastersoft]$ scp -r flink-1.10.2/ hadoop@slave2:/soft/
```

5．启动验证

```
[hadoop@mastersoft]$ flink-1.10.2/bin/start-cluster.sh
```

启动后，查看进程。

```
[hadoop@master soft]$ jps
2646 NameNode
2858 SecondaryNameNode
3310 StandaloneSessionClusterEntrypoint
3007 ResourceManager
[hadoop@slave1 soft]$ jps
2530 NodeManager
2681 TaskManagerRunner
2458 DataNode
[hadoop@slave2 soft]$ jps
2448 DataNode
2614 TaskManagerRunner
2520 NodeManager
```

可以看到在主节点 master 上有 StandaloneSessionClusterEntrypoint 进程，在从节点 slave1 和 slave2 上有 TaskManagerRunner 进程。还可以访问 master 节点的 Web 界面：http://192.168.128.131:8081/。

6．停止 Flink 集群

```
[hadoop@mastersoft]$ flink-1.10.2/bin/stop-cluster.sh
```